...n Fiji in 197~ ...~ grew up in ~~ ...~ ...~ ...~ UK.
~~ took history and politics at Edinburgh University,
...ards going on to edit the OneWorld.net website until
~~ He has made many appearances in the media as a
...mentator on environmental issues and is the author of
...*Tide: News from a Warming World* and the Collins Gem
...*n Calculator*. He was selected as a National Geographic
...ging Explorer in 2006. He lives in Oxford.

Join the debate at www.marklynas.org

...the reviews of *Six Degrees*:

...ntists predict that global temperatures will rise by
...en one and six degrees over the course of this century
...Mark Lynas paints a chilling, degree-by-degree picture
...~ devastation likely to ensue unless we act now ... *Six
...~es* is a rousing and vivid plea to choose a different
...~e' *Daily Mail*

...his book for everyone you know: if it makes them join
...ght to stop the seemingly inexorable six degrees of
...ning and mass death, it might just save their lives'
New Statesman

...~n with passion and packed with an impressive amount
...~rmation' *Guardian*

'In ...~ highly accessible book, Lynas lays out just what we
can expect with each progressive temperature rise, before

stating exactly what needs to happen regarding decreasing carbon emissions, among other things. This stuff used to be the preserve of scientists and governments. As Lynas makes painfully clear, it is now our problem, too' *Metro*

'An apocalyptic primer of what to expect as the world heats up ... it's sobering stuff and shaming too. Despite its sound scientific background, the book resembles one of those vivid medieval paintings depicting sinners getting their just deserts' *Financial Times*

'Lynas is interested in leaving the reader ready for action, rather than depressed ... a gripping storyteller' *Nature*

'These predictions are by no means far-fetched, based as they are on solid paleontological evidence ... There are no illustrations in this book but it doesn't need any, so vivid is the prose ... a most timely book ... the world's politicians and opinion formers should also read it and act on it' *International Journal of Meteorology*

By the same author

High Tide: News from a Warming World
Collins Gem *Carbon Calculator*

MARK LYNAS

SIX DEGREES

OUR FUTURE ON A HOTTER PLANET

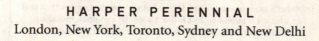

HARPER PERENNIAL
London, New York, Toronto, Sydney and New Delhi

Harper Perennial
An imprint of HarperCollins*Publishers*
1 London Bridge Street, London SE1 9GF

www.harperperennial.co.uk
Visit our authors' blog at www.fifthestate.co.uk

This updated edition published by Harper Perennial 2008

First published in Great Britain by Fourth Estate in 2007

A catalogue record for this book is available from the British Library

ISBN 978-0-00-720905-7

Set in Minion

Printed by CPI Group (UK) Ltd, Croydon CR0 4YY

MIX
Paper from
responsible sources
FSC® C007454

FSC is a non-profit international organisation established to promote the
responsible management of the world's forests. Products carrying the FSC
label are independently certified to assure consumers that they come
from forests that are managed to meet the social, economic and
ecological needs of present and future generations.

Find out more about HarperCollins and the environment at
www.harpercollins.co.uk/green

*To my wife, Maria, son, Tom, and daughter, Rosa,
in the hope that most of the predictions here
need not come true.*

CONTENTS

ACKNOWLEDGEMENTS ix

INTRODUCTION xiii

One Degree 1

Two Degrees 49

Three Degrees 99

Four Degrees 161

Five Degrees 191

Six Degrees 215

Choosing our Future 243

NOTES 281

INDEX 335

CONTENTS

ACKNOWLEDGEMENTS

INTRODUCTION

One Degree

Two Degrees

Three Degrees

Four Degrees

Five Degrees

Six Degrees

Choosing our future

NOTES

INDEX

ACKNOWLEDGEMENTS

This book is above all a work of synthesis, bringing together research conducted by many hundreds of scientists around the world. I cannot thank them all individually, but gratefully acknowledge that without their insight, expertise and dedication, *Six Degrees* would not exist, and we would be none the wiser about what lies ahead. I would also like to thank those who helped me locate and use the many and various sources on which this book depends: Jenny Colls at the Earth Sciences Library of Oxford University, and the staff of the Radcliffe Science Library in particular, where I spent many weeks and months buried in work down in the Lancaster Room. Unfortunately, not everything was always available – even given the resources of the Bodleian Library – so many thanks also to Jonathan Lifland of the American Geophysical Union for emailing me PDF copies of dozens of papers as published in *Geophysical Research Letters* and other AGU journals.

I am also indebted to my agent, Antony Harwood, another Oxford resident and frequent lunch companion, who instantly understood my Six Degrees idea from the very earliest moment. James Macdonald Lockhart at Antony Harwood Ltd also provided helpful advice on a number of occasions. Thanks in particular to my editor, Mitzi Angel at Fourth Estate, for taking the book on and supporting it (and me) so enthusiastically throughout the whole process. Silvia Crompton at Fourth Estate also put in a lot of work, as did copy-editor Merlin Cox, who improved notably upon my early draft. I should not forget proof-reader Anne Rieley, who thankfully picked up my silly mistakes and typos just in time.

Glaciologist and kindred spirit Stephan Harrison had a look at an early draft of *Six Degrees*, as did my old friend and squash partner Paul Kingsnorth – their comments were appreciated, even when ignored! Lastly, and most importantly of all, I am eternally grateful to my wife Maria for never once doubting me – even when the stresses and strains of writing (not to mention the subject matter) made me a less than ideal husband.

From the weeping ground there sprang a wind,
flaming with vermillion light,
which overmastered all my senses,
and I dropped like a man pulled down by sleep.

Dante, *Inferno*, Canto III:
Dante enters the First Circle of Hell

INTRODUCTION

The knock on the door came at night. In the darkness I could see two yellow jackets over black uniforms – the police. They were going door to door, the officers explained, to warn people in the area of the imminent risk of flooding. They handed over a photocopied leaflet, advising that we prepare to turn off the power and move all valuables upstairs, and were gone.

The rain had come two days earlier. It poured with torrential force for most of the day, accompanied by vivid flashes of lightning and intermittent claps of thunder. Roads were awash as flash floods swept off fields. Within hours, the rail link north was cut, and Oxford – like many other towns in the Midlands and southern England – was marooned. Four days later the waters were still rising, as a flood crest surged down the river Thames from more heavily inundated areas upstream. Turning on the television news I saw the pretty cathedral town of Tewkesbury turned into an island, both Cheltenham and Gloucester hit by power failures, and schools closed across the entire region. The rising flood swept over a water treatment plant, leaving a quarter of a million people with no drinking water for over a week. Though my own house was not flooded, whilst writing this I can still smell the stench of rotting waterweed left behind by the river on nearby Port Meadow.

The sheer intensity and violence of the rain reminded me of a tropical storm I rode out a few years earlier on the Outer Banks of North Carolina, whilst researching my first book *High Tide*. There was that same ominous dark quality to the sky, and the rainfall radar on the Meteorological Office website showed the same reds and whites of super-intense precipitation that I had previously witnessed whilst sheltering in the hurricane trackers' van near Cape Hatteras in 2002. Hurricanes generate some of the heaviest rainfall on Earth, and flooding during a hurricane strike is virtually a certainty. As the terrible drama that unfolded in New Orleans when Hurricane Katrina hit in 2005 showed, sometimes this flooding – combined with a monster storm surge – can be deadly.

All these events were windows into a changing world. Global warming is making the hydrological cycle more intense, causing heavier storms and more intense hurricanes to brew up out at sea. Yes, extreme weather has always been with us, but the fact that rising levels of greenhouse gases trap the sun's heat means that more energy is available in the system – so the worst is happening more and more often. The misery suffered in New Orleans three years ago felt to me like an insight into what the twenty-first century may have in store for many more of us, in a thousand locations across the world, as climate change accelerates.

The scenes lingered in my mind even as the city was emptied and the bedraggled survivors of New Orleans and the wider Gulf region were packed off to temporary shelters in Texas and elsewhere, where half a million still remain at the time of writing: arguably the first climate refugees, displaced permanently from their homes. I kept wondering: where next? What will happen as the world warms bit by bit? With up to six degrees Celsius of global warming on the cards over the next hundred years, according to the Intergovernmental Panel on Climate Change (IPCC), what will happen to our coasts, our towns, our forests, our rivers, our croplands and our mountains? Will we all, as some environmentalists suggest, be reduced to eking out a living from the shattered remains of civilisation in Arctic refuges, or will life go on much as before – if only a little warmer?

As I pondered these questions, I had already begun to sift through the latest scientific literature on global warming. I knew from earlier research for *High Tide* that scientists have now made hundreds of projections – mostly based on complex computer models – of how future global warming will affect everything from maize crops in Tanzania to snowfall in the Alps. Occasionally a particularly striking study makes headlines in the newspapers, but the vast majority of these forecasts are buried in obscure specialist journals, destined to be read only by other climatologists. Most of these journals are taken by Oxford University's Radcliffe Science Library, where they sit – undisturbed for weeks or even years on their dimly lit shelves – just a mile or so down the road from my own house. I realised that it was almost as if I had a Delphic Oracle in my back garden or Nostradamus living next door – except that these scientific prophecies were already coming true.

Earlier that year I had begun to make a daily pilgrimage down to the Radcliffe Science Library basement with my laptop, where as the weeks passed by I trawled through tens of thousands of scientific papers. Seasons came and went, and I barely noticed. Each relevant article, I slotted into a spreadsheet – papers about two degrees of global warming went into the two degrees slot, papers about five degrees of global warming went into the five degrees slot, and so on. Not all were computer model projections – some of the most interesting material came from palaeoclimate studies, investigations of how variations in temperatures have affected the planet during previous global warming events in prehistory. These records of past greenhouse episodes, I realised, could be analogues for the future: and they too slotted into my six degrees table according to the temperatures of the climatic periods they represented.

At the end, I found I had something truly unique: a degree-by-degree guide to our planet's future. And so, based on this raw material, the book gradually took shape: my first chapter included all the global warming impacts I could find associated with a one-degree rise in temperature, my second chapter covered two degrees, my third chapter covered three degrees … and on up the scale to six degrees – the worst-case scientific scenario. No scientist and no journalist has ever undertaken this work before with anything like this attention to detail, and never before has so much of this information been presented comprehensibly to a general audience in book form.

As the work emerged, I felt a nagging suspicion that maybe I should be keeping it all secret. *Six Degrees* was beginning to feel like a survival manual, full of indications about which parts of the globe might need to be abandoned, and which would be most likely to remain habitable. Maybe I should be sharing this information only with my family and friends, to give those people closest to me a quiet heads-up? Or perhaps I should get it out as widely as possible, as a sort of cautionary tale, to convince people to campaign for rapid emissions cuts and avoid the worst-case scenarios before it is too late?

Obviously I chose the second, more optimistic course. But a related question continued to bug me as I did early public presentations of *Six Degrees* material, particularly when I overheard a conversation in the toilets after one event in which an audience member apologised to another for dragging them out to something so depressing. I was truly shocked. Depressing? It had honestly never occurred to me that *Six Degrees* might be depressing. Yes, the impacts presented are terrifying – but they are also, in the main, still avoidable. Getting depressed about the situation now is like sitting inert in your living room and watching the kitchen catch fire, and then getting more and more miserable as the fire spreads throughout the house – rather than grabbing an extinguisher and dousing the flames.

It also dawned on me gradually when I tried to explain the book to non-specialists that most ordinary people have not got the slightest idea what two, four or six degrees of average warming actually means in reality. These still sound like very small changes when the mercury swings by fifteen degrees between night and day. To most of us, if Thursday is six degrees warmer than Wednesday, it doesn't mean the end of the world, it means we can leave the overcoat at home. Such are the vagaries of everyday weather. But six degrees of global average change is an entirely different prospect.

Consider this: 18,000 years ago, during the deepest freeze of the last ice age, global temperatures were about six degrees colder than today. In that frigid climate, ice sheets stretched across North America from sea to shining sea. As glacial grooves in the rocks in Central Park attest, New York was buried under a thick slab of ice, more than a mile deep as it stretched into the heart of the continent. Northern New Jersey was buried, as was all the Great Lakes area, and almost the entirety of Canada. Further south, the agricultural heartland of states like Missouri and Iowa would have been freezing tundra, blasted by dust-laden winds sweeping down from the ice cap, and underlain by layers of solid permafrost. During the ice age, humans were displaced far to the south, where places that are now subtropical, like Florida and California, maintained a temperate climate.

In addition, temperature swings were astonishingly rapid – several degrees in the space of a decade as the climate warmed and then cooled again. At one point, about 70,000 years ago, a huge supervolcano eruption in Indonesia blew thousands of cubic kilometres of dust and sulphur into the atmosphere, cutting off the Sun's heat and causing global temperatures to plummet. Humans were nearly wiped out in the ensuing 'nuclear' winter: the entire global human population crashed to somewhere between 15,000 and 40,000 individuals, a survival bottleneck which is still written in the genes of every human alive today. By implication, if six degrees of cooling was enough to nearly wipe us out in the past, might six degrees of warming have a similar effect in the future? That is the question this book seeks to answer.

Back in the summer of 2005, as I began my journey into humanity's likely future, I felt like Dante at the gates of the Inferno – privileged to see what few others have laid eyes upon, but also deeply worried by the horrors that seemed to lie ahead. Just as the poet Virgil was Dante's guide as he set forth into the Inferno, my guides are the many talented and passionate scientists who conducted the original research studies on which this book is based. I offer them my thanks, and hope they feel well represented by what follows.

> 'Set out then, for one will prompts us both.
> You are my leader, you my lord and master,'
> I said to him, and when he moved ahead
> I entered on the deep and savage way.

A technical note

As befits the task of any popular science writer, I have tried to make each case study come alive as much as possible without losing the rigour of the original document. Where the science itself has evolved through the years, I have tried to work this into the story. There were drawbacks of course: almost all the studies use different models, each model employing different underlying assumptions, so comparing them can sometimes be rather like comparing chalk and cheese. Each study also contains uncertainties, often expressed in quantitative terms – such is the nature of good science – and carefully weighed, thoughtful statements by the authors which cannot always be accurately reflected in a broad-brush, generalist approach such as this. I leave readers with queries about any of the information presented to follow up references and judge the original work for themselves. Do not complain to me either if you have doubts about the methodologies employed by the original studies: I am not a climatologist, I am merely the interpreter.

I might also add at this point, for the benefit of any readers who feel somewhat out of depth with the generally 'scienticised' nature of the climate change debate, a very general note of background on global warming. Essentially, this term (which I use interchange-ably with 'climate change', although technically they do mean slightly different things) refers to the increase in global atmos-pheric temperatures as a result of increasing concentrations of greenhouse gases in the air around us. That greenhouse gases have a warming effect, rather like an extra blanket around the globe, is indisputable, and has been established physics for over a hundred years. These gases cause a 'greenhouse effect' because they are opaque to long-wave infrared radiation: heat coming in from the Sun is short-wave, and so passes straight through, but when this heat is re-radiated by the Earth, its wavelength is longer, and some is trapped by the gases – just as glass in a greenhouse also traps heat. If there were no greenhouse gases at all in the atmosphere, the Earth's average temperature would be about −18°C.

Since the beginning of the Industrial Revolution, concentra-tions of the principal greenhouse gas, carbon dioxide (CO_2), have risen by a third, whilst those of methane – another potent green-house gas – have doubled. Although there have been fluctuations between the decades, global temperatures have also risen in the last 150 years by about 0.8°C, and are expected to rise even faster over the next century as CO_2 levels rise further still. Partly these future temperature rises will be the result of emissions already in the past, and partly they will reflect rapid expected rises in greenhouse gas emissions from human activity. That we can avoid higher temperature increases by cutting back emissions is a key point that I seek to illustrate in this book.

Although I have done my best to ensure that the correct impact studies are presented in the correct chapters, there are occasions when the decision about what to put where is somewhat arbitrary. Many – most, in fact – papers do not state the precise global average temperature change that their study refers to, particularly if they are focusing on a regional change. A study on Arctic sea ice, for example, may be based on a range of different future carbon dioxide concentrations, none of which are interpreted as global temperature averages by the authors, leaving me with the difficult choice of estimating which chapter is the best fit. Different studies using the same future CO_2 concentrations do not necessarily share the same temperature projections, moreover: all models have different 'sensitivities' to atmospheric greenhouse gas increases, further complicating the procedure. It is important to emphasise, however, that all of the material in this book comes from the peer-reviewed scientific literature – at no point do I base predictions on less reliable sources like newspaper articles or campaign group press releases.

It is also important to note that the temperature scale of this book is based on the IPCC's landmark 1.4 to 5.8°C temperature range, published in its 2001 Third Assessment Report, which gives us predictions of *up to* six degrees. This is reflected in the structure of the chapters that follow. The three degree chapter, for example, covers global temperatures of 2.1°C to 3°C, whereas the six degree chapter covers 5.1°C to 5.8°C. In February 2007 the IPCC published its Fourth Assessment Report (AR4), which broadened the range of temperature projections for 2100. For the lowest emission scenario, where global greenhouse gas emissions dip sharply, warming by 2100 could be as low as 1.1°C, according to the AR4, whereas for the highest emissions scenario, global warming could reach 6.4°C. In other words, the range is broader, and the worst-case scenario is even more drastic than in the 2001 IPCC report – seven degrees on this book's scale.

The Fourth Assessment Report of the IPCC also surveys in detail the expected impacts of future climate change, covering much of the same territory as this book and referencing many of the same papers. The language is sufficiently non-technical for most laypeople to find it perfectly comprehensible – something of an improvement on previous reports. I would in particular direct interested readers to the Working Group II section of the AR4, in particular a table in the Summary for Policymakers which outlines in a simple degree-by-degree scale the expected impacts of warming from 1 to 5°C. (Why the table does not extend to six degrees, despite this being within the temperature scenario projections given by the IPCC, is not explained.) The full text of all IPCC reports is available on the web at www.ipcc.ch.

An admitted pitfall in choosing a temperature-based structure for this book is that it makes giving dates very hazardous. The world could become two degrees warmer by 2100, for instance, or it could already have hit that level as early as 2030. The speed of warming is crucial in determining the capacity of human civilisation and natural ecosystems to adapt to the changing climate, and readers are urged to bear this in mind. The other option of running through the twenty-first century decade by decade would, I feel, have been even more problematic given that the dates attached to different emissions scenarios and temperature changes are highly uncertain. This book only deals with what scientists call 'transient' climate change: because of the thermal inertia of the oceans it will take centuries for temperatures to stabilise at any given concentration of greenhouse gases into a so-called 'equilibrium' state.

I have also on occasion explored rather speculatively what the changes projected by today's scientists might mean for society in future. Might China invade Siberia to secure subarctic *Lebensraum* in a globe where only narrowing zones remain habitable? Might India and Pakistan's struggle over the diminishing headwaters of Himalayan rivers turn nuclear as their people go thirsty? Of course, I would be foolish to expect these predictions to come true in any literal sense – history teaches us that human events are too unpredictable to support such a deterministic approach. But of this I have no doubt: climate change is the canvas on which the history of the twenty-first century will be painted. Forewarned is forearmed.

Onward, then. Let us enter the Inferno together.

1º

1°

1

ONE DEGREE

America's slumbering desert

It would be easy to walk right past them. Not many hikers pass this way, and those that do are unlikely to give a second thought to a few old stumps rooted in the river bed. In any case, this lonely spot, where the West Walker River canyon is at its narrowest as it plunges down the eastern flanks of California's Sierra Nevada, is not a place to linger – the area is notorious for sudden downpours and flash floods. The river runs almost the width of the entire gorge, and there's no place to climb to safety if the heavens open.

But these stumps have a story to tell. Dead trees can talk, in a way. An astute hiker or an observant angler would be puzzled: what are they doing in a river bed, a place now treeless because of the constant flowing water? Investigated by scientists in the early 1990s, the tree stumps were found to be Jeffrey pines – a common enough species for the area, but one that certainly doesn't normally root in rivers. What's more, these trees were old. Very old. Tissue samples revealed that the stumps dated from medieval times, and grew during two specific periods, centred on AD 1112 and 1350.

The mystery deepened when similar old stumps were revealed in Mono Lake, a large saltwater body a hundred kilometres to the south of Walker River, near the state border with Nevada. It's a

3

spectacular location, famous for broad skies and sunsets, with little to interrupt the gently rolling arid landscape other than a few extinct volcanoes. The Mono Lake tree stumps belonged not just to pines, but also to other native species like cottonwoods and sagebrush, all rooted far below current-day natural lake levels and only revealed thanks to water diversion projects that supply far-away Los Angeles. Again, carbon dating revealed the same two time intervals as for the Walker River trees. Clearly, something significant had happened back in medieval times.

More evidence came from deeper in the mountains, hidden in two locations today famous for their giant sequoia groves – Yosemite and Giant Sequoia National Parks. These enormous trees, which in terms of total wood volume stand as the largest living organisms on Earth, are also among the oldest. Some living trees are up to 3,000 years old. And because each annual growth cycle leaves a clear ring, these monumental plants are also an excellent record of past climate. Over a decade ago, scientists sampling wood from dead giant sequoias noticed old fire scars on the edges of some of their rings. These scars were especially frequent during this same medieval period – between AD 1000 and 1300 – as the old trees in West Walker River and Mono Lake were growing. Wildfires had raged in both national parks twice as frequently as before, and there can only be one plausible explanation – the woods were tinder-dry.

Raging wildfires, dry rivers and lakes – the pieces of the jigsaw were beginning to make sense. The area we now call California had in medieval times been hit by a mega-drought, lasting at different periods for several decades, and altering both landscape and ecosystems on a scale that dwarfs today's drought episodes. But just how geographically widespread was this event? Evidence from another lake, far away on the Great Plains of North Dakota, provides a partial answer. Moon Lake, like Mono Lake in California, is a closed basin, making it saline. Salinity fluctuates with the climate: in sequences of wet years, more fresh water ends up in the lake and salt levels go down. The converse is also true: in dry years, more water evaporates, leaving a more concentrated salty brine behind. Canadian scientists have now reconstructed long-term records of Moon Lake's saltiness by sampling the remains of tiny algae called diatoms – whose type and number fluctuate with salinity levels – from old lake sediments. Lo and behold, back before AD 1200, a series of epic droughts had swept the Great Plains, the return of which – the scientists agreed – 'would be devastating'.

An insight into the devastating nature of such a drought was gained by a team of biologists working in northern Yellowstone National Park, a good 1,500 kilometres to the south-west of Moon Lake, in Wyoming. They drilled into sediments spilled out by rivers, only to discover a peak in muddy debris flows – the product of flash floods – about 750 years ago. These flash floods had poured off mountainsides denuded of forest cover by frequent fires: so rather oddly, these flood debris flows are actually a classic sign of drought. It appeared that the whole of the western United States had been struck at the same time.

The effect on Native American populations in this pre-Columbian era was indeed devastating. Whole civilisations collapsed, beginning in the Chaco Canyon area of modern-day New Mexico. One of the most advanced societies on the continent at their peak, the Pueblo Indian inhabitants of Chaco Canyon erected the largest stone building on the North American continent before the European invasion, a 'great house' four storeys high, with over 600 individual rooms – much of it still standing today. Yet when the big drought came in AD 1130, they were vulnerable – population growth had already diminished the society's ecological base through the overuse of forests and agricultural land. Most people died, whilst the survivors went on to eke out a living in easily defended sites on the tops of steep cliffs. Several locations show evidence of violent conflict – including skulls with cut marks from scalping, skeletons with arrowheads inside the body cavity, and teeth marks from cannibalism.

Indeed, the whole world saw a changing climate in medieval times. The era is commonly termed the 'Medieval Warm Period', a time when – so the oft-told story goes – the Vikings colonised Greenland and vineyards flourished in the north of England. Temperatures in the North American interior may have been 1 to 2°C warmer than today, but the idea of a significantly warmer world in the Middle Ages is actually false. Recent research piecing together 'proxy data' evidence from corals, ice cores and tree rings across the northern hemisphere demonstrates a much more complicated picture, with the tropics even slightly cooler than now, and different regions warming and then cooling at different times. However small the global shift, the evidence is now overwhelming that what the western US suffered during this period was not a short-term rainfall deficit, but a full-scale mega-drought lasting many decades at least. As recently as 2007 US scientists reported tree-ring studies reconstructing medieval flows in the Colorado River at Lees Ferry, Arizona, showing that the river lost 15 per cent of its water during a major drought in the mid-1100s. For sixty years at a time, the river saw nothing but low flows – none of the floods that normally course down the Colorado arrived to break the dry spell. Indeed, the remarkable coincidence of these dates with evidence from New Mexico suggests that this was the very same drought that finished off the Chaco Canyon Indians.

To see the worst that even such a small change in climate can do, consider that most undramatic of places – Nebraska. This isn't a state that is high up on most tourists' 'to do' lists. 'Hell, I thought I was dead too. Turns out I was just in Nebraska,' says Gene

Hackman in the film *Unforgiven*. A dreary expanse of impossibly flat plains, Nebraska's main claim to fame is that it is the only American state to have a unicameral legislature. Nebraska is also apparently where the old West begins – local legend in the state capital Lincoln insists that the West begins precisely at the intersection of 13th and O Streets, a spot marked by a red brick star.

But perhaps the most important Nebraska fact is that it sits in the middle of one of the most productive agricultural systems on Earth. Beef and corn dominate the economy, and the Sand Hills region in central Nebraska sports some of the most successful cattle ranching areas in the entire United States.

To the casual visitor, the Sand Hills look green and grassy, and in pre-European times they supported tremendous herds of bison – hence their high productivity for modern-day beef. But, as their name suggests, scratch down a few centimetres and the shallow soil quickly gives way to something rather more ominous: sand. These innocuous-looking hills were once a desert, part of an immense system of sand dunes that spread across thousands of kilometres of the Great Plains, from Texas and Oklahoma in the south, right through Kansas, Colorado, Wyoming, North and South Dakota, to as far north as the Canadian prairie states of Saskatchewan and Manitoba. These sand dune systems are currently 'stabilised': covered by a protective layer of vegetation, so not even the strongest winds can shift them. But during the Medieval Warm Period, when temperatures in the Great Plains region may only have been slightly warmer than now, these deserts came alive – and began to march across a fertile landscape which today is a crucial food basket for humanity. This historical evidence indeed suggests that even tiny changes in temperature could tip this whole region back into a hyper-arid state.

People who remember the 1930s Dust Bowl might think they have seen the worst drought nature can offer. In the toughest Dust

7

Bowl years, between 1934 and 1940, millions of acres of Great Plains topsoil blew away in colossal dust storms. One, in May 1934, reached all the way to Chicago, dumping red snow on New England. Hundreds of thousands of people, including 85 per cent of Oklahoma's entire population, left the land and trekked west. All this took only an average 25 per cent reduction in rainfall – enough for ploughed farmland to blow away, but the giant dunes stayed put. What awoke the dunes from their long slumber nearly a thousand years ago was drought on an altogether different scale – with dramatically less rainfall, sustained over decades rather than just years.

In a world which is less than a degree warmer overall, the western United States could once again be plagued by perennial droughts – devastating agriculture and driving out human inhabitants on a scale far larger than the 1930s calamity. Although heavier irrigation might stave off the worst for a while, many of the largest aquifers of fossil water are already overexploited by industrialised agriculture and will not survive for long. As powerful dust and sandstorms turn day into night across thousands of miles of former prairie, farmsteads, roads and even entire towns will find themselves engulfed by blowing sand. New dunes will rise up in places where cattle once grazed and fields of corn once grew. For farmers, there may be little choice other than to abandon agriculture completely over millions of square kilometres of what was once highly productive agricultural land. Food prices internationally would rise, particularly if serious droughts hit other areas simultaneously. And although more southerly parts of the United States are expected to get wetter as the North American monsoon intensifies, residents here may not welcome an influx of several million new people.

Further east, however, agriculture may actually benefit from warmer temperatures and higher rainfall. Rather as California offered sanctuary of a sort to displaced 'Okies' during the Dust Bowl, the Midwest and Great Lakes areas will need to provide jobs and sustenance to those who can no longer scratch a living from the sandy soils far out west, once the rains stop falling and the desert winds begin to blow.

Already the day after tomorrow?

Just as farmers on the High Plains of North America are watching their fields and grasslands blowing away in the relentless heat, their kinfolk across the Atlantic may be grappling with another problem: extreme cold. One of the most counter-intuitive projected impacts of global warming is the possible plunging of temperatures throughout north-west Europe as the warm Atlantic current popularly known as the Gulf Stream stutters and slows down. This is the scenario fictionalised in an exaggerated form by the Hollywood disaster epic *The Day After Tomorrow*, where a collapse in the Atlantic current triggers a new ice age, flash-freezing New York and London (although the good guy still gets the girl). Real-world scientists were quick to lambast the film for flouting the laws of thermodynamics, but they also acknowledged that the reality of a slowdown in the North Atlantic Ocean current may still be pretty scary, especially for those who live in a part of the world which is used to a mild maritime climate far out of keeping with its high northern latitude.

A short technical aside is required here. Only a small part of the great current that delivers warm water into the North Atlantic is actually the real Gulf Stream: it, as its name suggests, is a stream of warm subtropical water heading north-east out of the Gulf of Mexico, which eventually becomes part of the much larger system of currents known to scientists as the Atlantic Meridional Overturning Circulation. The MOC is partly driven by the cooling and sinking of water at high latitudes off the coast of Greenland and Norway, where freezing Arctic air lowers its temperature and squeezes fresh water out as sea ice, leaving behind a heavy, salty brine which quickly sinks to the bottom of the ocean. From there it begins a return journey south – eventually surfacing (1,200 years later) in the Pacific. Scientists have long feared that a freshening and warming of the Norwegian and Greenland seas – due to higher rainfall, run-off from melting land glaciers and the disappearance of sea ice – could stop this water sinking, and shut down the great ocean conveyor. Hence the famous 'Shutdown of the Gulf Stream' scenarios familiar from newspaper headlines and the Hollywood movie.

Far-fetched it may seem, but Atlantic circulation shutdown has always been more than just a theory. It has happened before. At the end of the last ice age, 12,000 years ago, just as the world was warming up, temperatures suddenly plunged for over a thousand years. Glaciers expanded again, and newly established forests gave way once more to chilly tundra. The period is named the 'Younger Dryas', after an arctic-alpine flowering herb, *Dryas octopetala*, whose pollen is ubiquitous in peaty sediment layers dating from the time. In Norway temperatures were 7–9°C lower than today, and even southern Europe suffered a reversal to near-glacial conditions. On the other side of the Atlantic, cooling also occurred, and there is evidence of rapid climate change from as far afield as South America and New Zealand.

The culprit seems to be the sudden shutting-off of the Atlantic circulation due to the bursting of a natural dam holding back Lake Agassiz, a gigantic meltwater lake which had pooled up behind the retreating North American ice sheets. When the dam broke, an enormous surge of water (the lake's volume was equivalent to seven times today's Great Lakes) is thought to have poured through Hudson Bay and out into the Atlantic. This freshwater surge diluted the North Atlantic seas and stopped them being salty enough to sink, interrupting the deep ocean current and triggering climatic destabilisation across the world.

Obviously today there are no gigantic ice lakes waiting to flood into the North Atlantic, but global warming could still interrupt the formation of deep water by melting sea ice and causing greater freshwater run-off from Siberian rivers. Despite the rapidly melting ice cap, however, for many years there was no evidence that changes in the Atlantic MOC were actually happening, and many oceanographers had begun to pooh-pooh the theory. That was until the RSS *Discovery*, a scientific research vessel owned by the British government, began a routine cruise across the Atlantic in 2004. The ship's scientific team set themselves the task of sampling seawater at various depths on a line drawn between the Canary Islands in the east and Florida in the west, aiming to repeat similar measurements taken in 1957, 1981, 1992 and 1998. They had not expected to discover anything terribly exciting; in fact the team leader Professor Harry Bryden confided to one journalist: 'In 1998 we saw only very small changes. I was about to give up on the problem.'

But 2004 was different. Bryden and his colleagues found that less warm water was flowing north at the surface and less cold water was flowing south at depth. Overall, the Atlantic circulation had dropped by 30 per cent, equivalent to the loss of 6 million tonnes of water flow per *second*. No wonder Professor Bryden admitted that he was 'surprised'. Suddenly the slowing-down of the great Atlantic current system was no longer just a hypothesis postulated for the distant future. It was already happening.

The media reaction was instantaneous. 'Current that warms Europe weakening', warned CNN. NPR's *All Things Considered* show led with 'Atlantic Ocean's heat engine chills down'. In Europe, the response was one of understandable concern. 'Alarm over dramatic weakening of Gulf Stream', reported the UK's *Guardian* newspaper on 1 December 2005. 'Global warming will bring cooler climate for the UK' was the *Telegraph*'s take on the same story. A couple of paragraphs down, the paper reported one expert as confirming that 'an average temperature drop of a degree or two within decades would herald more extreme winters'.

Older readers would have shuddered at the thought of a return to winters as bitter as that of 1962–3, when the UK was blanketed in snow for more than three months, and temperatures hit a low of −16°C in southern England. In places the sea froze, and ice floes appeared in the river Thames at London's Tower Bridge. That season was about 2.7°C colder than average – almost exactly the temperature drop predicted for London in one modelling study investigating the possible result of a 50 per cent drop in the warm Atlantic current. Was Europe's new ice age just around the corner?

Apparently not. Almost exactly a year later, and with much less fanfare, *Science* magazine reported that 'a closer look at the Atlantic Ocean's currents has confirmed what many oceanographers suspected all along: there's no sign that the ocean's heat-laden "conveyor" is slowing'. Instead of just the snapshot data generated by a few irregular ship cruises, nineteen permanent instrument-laden sensors had now been stretched across the Atlantic between West Africa and the Bahamas – and they were able to deliver a much more consistent picture. A year of continuous monitoring, Harry Bryden now reported to a conference in Birmingham, showed that his original 30 per cent decline was just a part of random natural variability after all, the sort of thing that happens constantly from one year to the next.

This result was a triumph for the modellers, most of whom had for years been pouring cold water on the European ice age theory. They agreed that huge volumes of freshwater would need to surge into the North Atlantic in order to shut off the Gulf Stream – far more than currently being generated by melt from Greenland or higher precipitation in Siberia. Rather than plunging overnight, the ocean circulation might decline by a stately 25 to 30 per cent or so, but only after at least 100 years of sustained greenhouse gas emissions. Even then, it wouldn't cool Europe – it would simply moderate the otherwise rapid rise in temperatures.

As the IPCC concluded in 2007: 'it is … very unlikely that the MOC [Atlantic Meridional Overturning Circulation] will undergo a large abrupt transition during the course of the 21st century'. Although all of them showed some weakening by 2100, none of the models assessed by the IPCC supported the collapse scenario. And even with this MOC slowdown, the IPCC reported that 'there is still warming of surface temperatures around the North Atlantic Ocean and Europe due to the much larger effects of the increase in greenhouse gases'. The IPCC's judgement was final: there would be no new ice age for Europe.

Africa's shining mountain

The amateur adventurer Dr Vince Keipper had waited years for this day. Nearing the summit of Kilimanjaro, the highest point on the African continent, Keipper and his group were looking forward to panoramic views of the surrounding Kenyan and Tanzanian plains. They had climbed through the steep and treacherous Western Breach and past the towering ice cliffs of the Furtwängler Glacier. The weather was perfect, with only a few clouds far beneath. Then, not far from the top of the 5,895-metre peak, a loud rumbling sound from behind them brought the group to a sudden halt. 'We turned around to see the ice mass collapse with a roar,' remembered Keipper. 'A section of the glacier crumbled in the middle, and chunks of ice as big as rooms spilled out on the crater floor.'

Keipper and his group knew they had had a lucky escape: they might have been buried had the collapse happened only a few hours earlier. They also knew that the event they had just witnessed had a powerful symbolic resonance: right in front of their eyes, the highest peak in Africa was melting.

Kilimanjaro has become something of a poster child for the international climate change campaign. The Swahili words *kilima* and *njaro* translate as 'shining mountain', testament to the power of this massive volcano to inspire awe in onlookers through the ages. A recent aerial photo of the crater, with little more than a few ice fragments encrusting its dark sides, was the centrepiece for a touring global warming photography exhibition sponsored by the British Council in 2005. During the 2001 UN climate change conference in Marrakech, Morocco, Greenpeace sent a team to Kilimanjaro to hold a press conference by video link from beside one of the mountain's disappearing glaciers. Kilimanjaro's international celebrity status has also attracted the attention of climate change deniers, who suggest that deforestation on the mountain's lower slopes is more to blame for glacial retreat than global warming.

None of the contrarian rhetoric cuts any ice, so to speak, with Lonnie Thompson, a glaciologist at Ohio State University and a man who is deservedly one of America's most celebrated natural scientists. Thompson pioneered the drilling of ice cores in inaccessible mountain regions, bringing back ice tens of thousands of years old from glaciated peaks as remote and far apart as Peru's Nevado Huascarán and Tibet's Dasuopu, often pushing himself to the edge of human endurance in the process. In 1993 Thompson and his drilling team camped for 53 days at 6,000 metres between the two peaks of Huascarán, perhaps setting a world record for high-altitude living. (I stayed there for one night in 2002 – one of the most freezing, wind-blasted and wretched nights of my life.)

At one point a gale blew Thompson's tent, with him inside, towards a precipice – until he jammed his ice axe through the floor. 'I don't understand,' he once remarked, 'why anyone would want to climb a mountain for fun.'

As Thompson was one of the first to recognise, this mountain ice contains a unique record of climate variations down the ages – preserved in layers of dust, isotopes of oxygen and tiny bubbles of gas trapped within the frozen layers of water. Once carried down in freezer boxes and analysed in the laboratory, these icy signatures trace everything from droughts to volcanic eruptions from decades and centuries past. They also tell a story about past temperature changes: the two isotopes of oxygen, ^{16}O and ^{18}O (which have different atomic weights due to the presence of two more neutrons in the latter's nucleus), vary in abundance with water temperature, so their proportions in ice cores are a good 'proxy' record of ancient climates.

Thompson and his team also drilled on three of Kilimanjaro's remaining glaciated areas, and in October 2002 concluded that 80 per cent of the mountain's ice had already melted during the past century. The news made international headlines, along with Thompson's prediction that the rest of the ice would be gone by between 2015 and 2020. As he readily admitted, this prediction was not based on complex computer modelling or any other advanced techniques. 'In 1912 there were 12.1 square kilometres of ice on the mountain,' he told journalists from CNN. 'When we photographed the mountain in February of 2000, we were down to 2.2 square kilometres. If you look at the area of decrease, it's linear. And you just project that into the future, sometime around 2015 the ice will disappear off Kilimanjaro.'

If there was an urgency in Thompson's voice, this was because he knew that recent melting had already begun to destroy the unique record of past climate preserved in Kilimanjaro's glaciers.

In their analysis of dust layers in the ice, the scientific team found evidence of a marked 300-year drought four thousand years ago; a drying so severe that it has been linked to the collapse of several Old World civilisations across North Africa and the Middle East. The ice also indicated much wetter conditions even longer ago, when huge lakes washed over what is now Africa's dry Sahel. Close to the surface Thompson's team discovered ice containing a layer of the radionuclide chlorine-36, fallout from the American 'Ivy' thermonuclear bomb test on Eniwetok Atoll in 1952. With this precise time control, the scientists could tell that ice which would have preserved a record of climate fluctuations since the 1960s had already melted away.

Moreover, the oldest ice at the base of the cores proved to be over 11,000 years old, showing that at no time since the last glacial epoch has the peak of Kilimanjaro been free of ice. This discovery made Thompson's ice cores even more valuable, for the simple reason that within as little as ten years the sawn-up circular cores in Ohio State University's walk-in freezer will be the only Kilimanjaro ice left anywhere in the world. With this in mind, Thompson and his team have already decided that some of the ice will be kept intact for future generations of scientists to dissect with new technologies, possibly unlocking climatic secrets still undreamt of today.

The efforts of climate change deniers to suggest that there is something special about the disappearance of Kilimanjaro's glaciers are undermined by similar changes taking place in mountain ranges right across the world, not least in the Rwenzori Mountains of Uganda, nearly a thousand kilometres to the north-west. In this remote region, where Uganda borders the Democratic Republic of the Congo, the fabled 'Mountains of the Moon' generate such heavy rainfall (about 5 metres per year) that the cloud-shrouded peaks are only visible on a few days out of every year, and form the main headwaters of the river Nile. At the top of the highest peak, the 5,109-metre Mount Stanley (named after the explorer, who passed by in 1887), ice and snow deny the summit to all but the most determined mountaineers. Yet as at Kilimanjaro, glacial retreat in the Rwenzoris has been profound: the three highest peaks have lost half their glacial area since 1987, and all the glaciers are expected to be gone within the next two decades.

Elsewhere in the world, disappearing mountain glaciers pose a major threat to downstream water supplies. But Kilimanjaro's ice cap is so small that its final disappearance will make little difference to the two major rivers – the Pangani and the Galana – which rise on its flanks. Instead, the crucial water link for Kilimanjaro is not the glaciers, but the forests. The montane forest belt at between 1,600 and 3,100 metres provides 96 per cent of the water coming from the mountain – this lush tangle of trees, ferns and shrubs not only captures Kilimanjaro's torrential rainfall like a giant sponge, but also traps moisture from the clouds which drape themselves almost permanently around the mountain's middle slopes. Much of this water drains underground through porous volcanic ash and lavas, and emerges in waterholes – vital for local people as well as for wild animals – far away on the savannah plains.

So is Kilimanjaro's water-generating capacity safe from global warming? Not quite: rising temperatures and diminishing rainfall increase the risk of fires, which have already begun to consume the upper reaches of montane forest. By the time the glaciers have disappeared, so will the higher forests, depriving downstream rivers of 15 million cubic metres of run-off every year, according to one estimate. In contrast, the loss of glacial water input will likely add up to less than 1 million cubic metres annually: significant, but not catastrophic. The diminishing water supply will affect everything from fish stocks to hydroelectric production downriver in poverty-stricken Tanzania. Much of the mountain's world-famous biodiversity (Kilimanjaro hosts twenty-four different species of antelope alone) will also be threatened by the weather changes.

As the snows disappear, so will much of the wildlife and the verdant forests that tourists currently trek through on their arduous journey to the roof of the African continent.

Ghost rivers of the Sahara

Far to the north of Kilimanjaro, in the Sahel, another drought-stricken area could by this time be experiencing some blessed relief. The Sahelian region of North Africa has long been synonymous with climatic disaster: during the 1970s and 80s famines struck the area with such severity that they sparked massive humanitarian relief efforts like Band Aid and Live Aid. Reporting from Ethiopia's refugee camps in 1984, the BBC's Michael Buerk spoke of a 'biblical famine' as the camera swept slowly over the dead and dying. Over 300,000 people perished during earlier famines in the 1970s.

The Sahel is an immense area, stretching in a wide belt east to west across northern Africa from Senegal on the Atlantic coast to Somalia on the Indian Ocean. For the most part savannah and thorn scrub, it is a climatic transition zone between the hyper-arid Sahara to the north and the lush tropical forests which grow nearer to the equator in the south. Intermittent rains mean that nomadic cattle herding has long been a dominant way of life, with people wandering far and wide through the seasons in search of grazing for their livestock. It is often assumed that global warming will further desiccate the Sahel, allowing the Saharan dunes to march south into Nigeria and Ghana, and displacing millions in the process. Although the forecasts are tentative and uncertain, both palaeoclimatic studies and computer models suggest that the reverse might be true. As other parts of Africa shrivel in the heat, could the Sahel end up as a refuge?

For clues to how the area's climate might alter, we need to venture north into the great Sahara. Here, the world's largest desert has also seen the highest temperature ever recorded on Earth: a truly blistering 58°C. The Sahara covers an area so huge that the entire contiguous United States would comfortably fit inside. This desert doesn't just have sand dunes, it has sand mountains, some reaching to nearly 400 metres in height. It is so completely uninhabitable that only a sprinkling of people get by in a few dwindling oases and at the desert's edge.

But scattered over this enormous area are clear signs that a very different Sahara existed many thousands of years ago. Neolithic paintings and rock carvings have been discovered in places where settled human existence is utterly impossible today. This ancient art depicts elephants, rhinoceroses, giraffes, gazelles and even buffalo – all animals which currently roam only hundreds of kilometres to the south. In Egypt's hyper-arid Western Desert, where less than 5 mm of rain falls on average each year, arrowheads and flint knives for hunting and butchering big game have been unearthed by archaeologists. At one site in south-western Libya, archaeologists even discovered tiny flint fish-hooks – again in an area where no trace of surface water persists now.

Other indications of a wetter past have also been discovered. Although anyone crossing Egypt's dry Safsaf Oasis by camel would today see little more than rock and dunes, radar pictures taken from the space shuttle *Endeavour* in 1994 clearly show whole river valleys buried beneath the sands. These ghostly watercourses even include major tributaries to the Nile flowing out through modern-day Sudan, all long-dry and forgotten beneath the dust. In southern Algeria, huge shallow lakes once gathered, supporting plentiful populations of fish, birds and even Nile crocodiles. The carbon dating of freshwater snails and desiccated vegetation preserved in these old lake beds shows that between five and ten thousand years ago the desert edge retreated 500 kilometres further north, and at different times almost disappeared altogether.

On the borders of what is today Chad, Nigeria and Cameroon, an immense lake, over 350,000 square kilometres in area, extended across the southern Sahara. Nicknamed Lake Mega-Chad, after its modern-day remnant Lake Chad, this gigantic inland sea was the largest freshwater body that Africa had seen for the last two and a half million years. It would have been only slightly smaller than today's largest lake, the Caspian Sea. Strange ridges of sand, which today lie marooned far away in the desert, reveal the shores of the old lake, as do the shells of long-dead molluscs which once thrived in its warm, shallow waters. The flat landscape between the marching dunes testifies to the erosive power of its long-vanished waves.

Common sense suggests that a major lake in such an arid area could only have been maintained by much higher rainfall, and longer-term records do indeed show that the Saharan region has experienced repeated wet and dry episodes over cycles of many thousands of years. The coldest periods of the ice ages tended to be the driest in the Sahara, whilst warm interglacials brought rain – allowing life to emerge once again. During the early Holocene epoch, 9,000 to 6,000 years ago, the northern hemisphere summer sun was slightly stronger than today, thanks to a small cyclical shift in the Earth's orbit around the sun. The increased heating warmed up the giant North African landmass to such an extent that it powered a monsoon – just like the one that brings annual summer rains to the Indian subcontinent today.

Monsoons are based on the simple principle that land surfaces heat up quicker in the summer than the surrounding oceans. This creates an area of low pressure as the hot air in the continental interior rises, sucking in cooler, moister air from the neighbouring seas. These rain-bearing winds bring torrential summer downpours to monsoonal climates such as India's, where agricultural life revolves with this annual cycle. The African summer monsoon is weaker and less generally recognised, but is still the only source of reliable rainfall for the Sahel. Climate models project that land surfaces will warm much faster than the oceans during the twenty-first century, potentially adding a boost to summer monsoons. So with one degree of global warming, this monsoon could begin to gain power and penetrate once again far into the African continent, greening the Sahara.

But will it actually happen? Before anyone makes plans to move large-scale food production to the central Sahara, a note of caution needs to be sounded. During the early Holocene, an additional monsoon driver was the difference in the distribution of solar heat between the two hemispheres. This time the whole globe is heating up, so the past is not a perfect analogue for the future. Moreover, it would be wrong to get the impression that the more humid Sahara was some kind of verdant wonderland – rainfall totals mostly only reached 100 mm or so, enough to support only the barest savannah-type vegetation, and wetter phases would also have been interspersed with long droughts. However, computer models can help negotiate a way through the conflicting possibilities – and the answer they provide holds profound implications for all the inhabitants of North Africa.

The preliminary stage is set by Martin Hoerling and two other climate scientists based in Boulder, Colorado, who used sixty different model runs to confirm that whilst southern Africa dries out with global warming, northern Africa does indeed begin to get wetter. Indeed, the long-term drying trend which caused such misery and devastation during the second half of the twentieth century goes into full reversal after about 2020 (with one-degree global warming or less), when the Sahel sees a long-term recovery in its rainfall. By 2050 the recovery is in full swing, with 10 per cent more rainfall right across the sub-Saharan zone.

This conclusion is supported by a second study, which projects heavier rains on both the West African coast and into the Sahel as a warmer tropical Atlantic Ocean supplies huge amounts of water vapour to form rain-bearing clouds. With more plentiful rains, crop production can potentially increase, offsetting declines else-where – assuming, that is, that temperatures are not so high that people who once died from famine now die from heatstroke.

However, computer modellers based in Princeton, New Jersey, have come up with a rather different long-range forecast. Their model accurately simulates the terrible 1970s and 1980s drought – but after a short interlude of higher rainfall, it projects even fiercer drought conditions for the Sahel region in the second half of the twenty-first century.

So why the divergence between the different models produced by the Princeton and the Boulder teams? The Princeton researchers admit that they are stumped. 'Until we better understand which aspects of the models account for the different responses in this region,' they caution, 'we advise against basing assessments of future climate change in the Sahel on the results of any single model.' Nevertheless, they insist, 'a dramatic 21st century drying trend should be considered seriously as a possible future scenario'.

This latter finding also chimes with global studies, which suggest stronger droughts affecting ever-larger areas as the world warms up. One of the most wide-ranging analyses was undertaken by Eleanor Burke and colleagues from the Hadley Centre at Britain's Meteorological Office, who used a measure known as the 'Palmer Drought Severity Index' to forecast the likely incidence of drought over the century to come. The results were deeply troubling. The incidence of moderate drought doubled by 2100 – but worst of all, the figure for extreme drought (currently 3 per cent of the planet's land surface) rose to 30 per cent. In essence, a third of the land surface of the globe would be largely devoid of fresh water and therefore no longer habitable to humans.

Although these figures are based on global warming rates of higher than one degree by 2100, they do indicate the likely direction of change. As the land surface heats up, it dries out because of faster evaporation. Vegetation shrivels, and when heavy rainfall does arrive, it simply washes away what remains of the topsoil. It may seem strange that floods and droughts can be forecast to affect the same areas, but with a higher proportion of rainfall coming in heavier bursts, longer dry spells will affect the land in between. This, then, is the most likely forecast for the Sahel: whilst rainfall totals overall may indeed rise, these increases will come in damaging flash-flood rainfall, interspersed with periods of intensely hot drought conditions.

According to some historians, the greener Sahara of 6,000 years ago was the geographical basis for the mythical Garden of Eden, its original inhabitants expelled not by God for bad behaviour, but by a devastating drying of the climate. Whilst scientists continue to argue over the specifics of the likely climatic future of the Sahara and Sahel, one thing seems clear: humanity will not be returning to Eden any time soon.

The Arctic meltdown begins

Over recent years a new phrase has entered the scientific lexicon: 'the tipping point'. Originally popularised by Malcolm Gladwell's bestselling book of the same name, the understanding that social or natural systems can be non-linear is a crucially important one. An oft-used analogy is of a canoe on a lake: wobble it a little, and stability can return with the boat still upright. Cross the point of no return, the 'tipping point', and the boat will capsize and find a new stability – this time upside down, with the ill-advised canoeist floundering underneath.

Scientists have increasingly realised that the Earth's climate is a good example of a non-linear system: over the ages it has been stable in many different states, some much hotter or colder than today. During the ice ages, for example, global temperatures averaged five degrees cooler than now for tens of millennia. Moreover, the system can 'tip' from one state into another with surprising rapidity. Episodic sudden warmings embedded within the last ice age saw temperatures in Greenland rise by as much as 16°C within just a few decades. The reasons why the climate flipped so rapidly are still not completely understood, but it is clear that even tiny changes in 'forcings' – from greenhouse gases or the Sun's heat – have in the past led to dramatic responses in the climate system. In contrast, our relatively stable climate is highly unusual – the Holocene period, during which all of human civilisation has come about, has seen very little change in global temperatures. Until now.

Scientists have established beyond reasonable doubt that the current episode of global warming, of about 0.7°C in the last century, has pushed Earth temperatures up to levels unprecedented in recent history. The IPCC's 2007 report confirmed that no 'proxy records' of temperature – whether from tree rings, ice cores, coral bands or other sources – show any time in the last 1,300 years that was as warm as now. Indeed, records from the deep sea suggest that temperatures are now within a degree of their highest levels for no less than a million years.

The part of the globe most vulnerable to this sudden onset of warming, and the part which will likely see the first important 'tipping point' crossed, is the Arctic. Here, temperatures are currently rising at twice the global rate. Alaska and Siberia are heating up particularly rapidly; in these regions the mercury has already risen by 2–3°C within the last fifty years.

The impacts of this change are already profound. In Barrow, Alaska, snowmelt now occurs ten days earlier on average than in the 1950s, and shrubs have begun to sprout on the barren, mossy tundra. Scientists based in Fairbanks, Alaska, have documented a sudden thawing of underground ice wedges on the state's normally cold North Slope, with new meltwater ponds dotting the landscape. These ice masses had previously remained frozen for at least the past three thousand years, indicating how far outside previous historical variability current warming is moving.

In other parts of the state entire lakes are draining away into cracks in the ground as the impermeable permafrost layer thaws underneath them. More than 10,000 lakes have shrunk or disappeared altogether in the last half-century, highlighting an alarming drop in the state's water table. In 2007, Canadian researchers reported that in Ellesmere Island, Nunavut, ponds which existed for millennia have now become ephemeral as their water evaporates away in the summer heat. Water-dependent species from insect larvae and freshwater shrimps to nesting birds are being wiped out as a result. Vegetation that once grew on these thin, waterlogged soils is now so desiccated that it easily catches fire.

Arctic mountain glaciers are also responding. On the Seward Peninsula of Alaska, the Grand Union Glacier is retreating so quickly that it is projected to disappear entirely by the year 2035. Other, much larger glaciers elsewhere in Alaska are also thinning rapidly. In the decade up to 2001 alone, the biggest Alaskan glaciers are estimated to have lost 96 cubic kilometres of ice, raising global sea levels by nearly 3 mm. Across the entire Arctic, glaciers and ice caps have lost 400 cubic kilometres of volume over the past forty years.

Perhaps the clearest bell-wether of change is found out at sea. The Arctic ice cap has been in constant retreat since about 1980, with each successive summer seeing more and more of its once-permanent ice disappearing. Each year on average 100,000 square kilometres of new open ocean is revealed as the ice which once overlay it melts away. In September 2005 alone, an area of Arctic sea ice the size of Alaska vanished without trace. Even in the pitch blackness of the winter months, the sea ice cover has been ebbing – both 2005 and 2006 saw the ice extent fall far below average.

Here is where the tipping point comes in. Whilst bright white, snow-covered ice reflects more than 80 per cent of the Sun's heat that falls on it, the darker open ocean can absorb up to 95 per cent of incoming solar radiation. Once sea ice begins to melt, in other words, the process quickly becomes self-reinforcing: more ocean surface is revealed, absorbing solar heat, raising temperatures and making it more difficult for the ice to re-form during the next winter. Climate models differ about exactly where the Arctic sea ice tipping point may lie, but virtually all of them agree that once we are past a certain threshold of warming the disappearance of the entire northern polar ice cap is pretty much unavoidable.

These models suggest that we have not yet reached this critical tipping point – but it may not lie very far away. One model run projects a sudden collapse in sea ice cover after 2024, with four million square kilometres of ice melting away in the following ten years. In this simulation, reported by a US-based team led by Marika Holland of the National Center for Atmospheric Research in Boulder, Colorado, the whole ocean becomes virtually ice-free in summertime by 2040. Whilst other model runs examined by the same team don't cross the tipping point until 2030 or 2040, one simulates a collapse in sea ice production beginning as early as 2012.

Even so, Holland's team emphasises that 'reductions in future greenhouse gas emissions reduce the likelihood and severity of such events' – in other words, all is not yet lost. Another team, led by NASA's Jim Hansen, reaches a similar conclusion. Despite major changes already in the system, Hansen and co-authors write, 'it may still be possible to save the Arctic from complete loss of ice' – but only if other atmospheric pollutants (such as soot, which darkens the ice surface and speeds melting) are reduced as well as carbon dioxide. Implement a dramatic programme of emissions reductions, and we 'may just have a chance of avoiding disastrous climate change', the team concludes. We may not have much time left, however: at the time of writing, 10 August 2007, a new historic sea ice minimum has just been reported for the Arctic. With a whole month of summer melting still left to go, the expectation is that the previous record low, recorded in 2005, will be 'annihilated'. Particularly worrying is that dramatic ice extent reductions are being recorded for every sector of the Arctic basin, whereas in previous years only certain areas were affected. Perhaps this is what a tipping point looks like.

But why is Arctic sea ice so important? As the following chapter will show, without it emblematic Arctic species like polar bears and seals are doomed to extinction. But the impacts will also hit closer to home, far away from the once-frozen north. As Ted Scambos, lead scientist at the US National Snow and Ice Data Center in Colorado, explains: 'Without the ice cover over the Arctic Ocean we have to expect big changes in the Earth's weather.'

These big changes are inevitable because of how the world's climate works. Most mid-latitudinal weather is generated by the contrast between polar cold and equatorial heat: the reason the UK gets year-round rainfall is because of its location on this unstable boundary between these competing air masses – the so-called 'polar front'. The nor'easter storms which barrel up the eastern US coastline in winter are also generated by this temperature contrast. But with the Arctic warming up, this contrast will lessen and the zone where it takes place will migrate north as rising temperatures contract the world's weather belts towards the poles. In the UK places like Cornwall and Wales which are accustomed to bearing the brunt of stormy winter weather may find themselves in the doldrums for weeks and months at a time, with a much drier overall climate. Only Scotland is likely to hang on to the wetter weather indefinitely. And as chapter 3 will show, the result in the western US is also likely to be drought – but on a scale never before experienced in human history.

Nor are these predicted changes just conjecture: they are already under way. Satellite measurements over the past 30 years have shown a marked 1° latitudinal contraction of the jet streams towards the poles in both hemispheres. Given that these high-altitude wind belts – narrow corridors of rapidly moving air at the top of the troposphere – mark the boundaries between the different air masses, their gradual movement shows that the location of the world's typical climate zones is already starting to shift in response to rising global temperatures.

What we have so far witnessed is still only the beginning. As one group of scientists warned recently: 'The Arctic system is moving toward a new state that falls outside the envelope of recent Earth history.' As future chapters show, this new ice-free Arctic will see extreme levels of warmth unlike anything experienced by the northern polar regions for millions of years.

Danger in the Alps

When the Englishmen Craig Higgins and Victor Saunders left the Hornli hut at 4 a.m. on 15 July 2003, they had no idea that they would end the day being part of the biggest-ever rescue on Switzerland's iconic Matterhorn. The ascent began straight-forwardly, with the two climbers scaling three rock towers, after which steep slabs led up to a small bivouac hut midway up the Hornli ridge. Higgins and Saunders had just reached the second hut, at 6 a.m., when an enormous rock avalanche pounded down the eastern face of the mountain. Cowering behind the building as stones bounced all around them, the two climbers would have been well advised at that point to turn tail and descend as quickly as possible. But mountains have strange effects on people's minds, and the two Brits pressed on.

Then, three hours later, the mountain shook once again as a further gigantic rockfall crashed down, this time from the north face. Shortly after, a third rockfall struck – and this time the Hornli ridge itself was giving way. A Swiss mountain guide found himself inches from disaster as the ground began to crumble just in front of him. With no hope of crossing the dangerously unstable zone, the guide radioed for help. For the next four hours two Air Zermatt helicopters ferried stranded climbers off the ridge and back to the main hut. 'As we climbed slowly down,' recalled Saunders, 'the smoking plume of rock dust and the returning helicopters told us of a major rescue taking place.' Both British climbers, realising they too were trapped, joined the queue of people waiting to be plucked to safety.

Ninety people were rescued that day, and amazingly no lives were lost or injuries reported – a tribute to the professionalism of the Swiss mountain guides and emergency services. The mountain remained closed for days afterwards as experts tried to assess the likelihood of further rockfalls taking place. In fact, falling rocks were not the only hazard in the area: on the same day as the Matterhorn drama was taking place, massive chunks of ice broke off from a glacier above the nearby resort of Grindelwald and plunged into a river, causing a two-metre-high wave to flood down the mountain. Fast-acting police managed to clear the area of holidaymakers just before the mass of rocks and mud washed by.

When he heard about the two near-disasters, the glaciologist Wilfried Haeberli had no doubts about the cause. 'The Matterhorn relies on permafrost to stay together,' the Zurich University scientist told reporters. But Switzerland had just been suffering its strongest-ever heatwave. With the fierce summer heat having melted all the winter snow much earlier than usual, the permafrost and glaciers themselves were beginning to melt down. Once that process begins, Haeberli warned, 'water starts to flow, and large chunks of rock begin to break away from the mountain'.

Most ground in the Alps above about 3,000 metres remains permanently frozen throughout the year, and is anchored, as Haeberli says, by permafrost. But in the summer of 2003 the melt zone reached as high as 4,600 metres – higher than the summit of the Matterhorn, and nearly as high as the top of Mont Blanc, western Europe's highest mountain. And whilst the Matterhorn climbers were lucky to get down safely on 15 July, at least fifty other climbers were less fortunate during that boiling summer – most were killed by falling rocks.

Haeberli, a world expert on permafrost, has since co-written a scientific paper on the impacts of the 2003 hot summer in the Alps. He and colleagues calculated that the thaw experienced during that heatwave outranked anything the mountains had suffered in recent history, and that most rock fall as a result took place during the hottest months of June, July and August.

They also found that the 2003 thaw penetrated up to half a metre deeper into the rock than any thaw during the previous two decades.

Surprisingly, however, the worst rockfalls didn't take place on sunny slopes where the direct heat was strongest, but on shaded northern faces, where the high air temperature penetrated the mountain. Ominously, the study concludes that with one degree of further global warming, more permafrost degradation in the Alps is unavoidable. 'Widespread rockfall and geotechnical problems with human infrastructure are likely to be recurrent consequences of warming permafrost in rock walls due to predicted climatic changes,' Haeberli and his colleagues warn. 'The extreme summer of 2003 and its impact on mountain permafrost may be seen as a first manifestation of these projections.'

As mountain slopes thaw out and fail, whole towns and villages will be at risk of destruction in the Alps and other mountain regions. Some towns, like Pontresina in eastern Switzerland, have already begun building earthen bulwarks to guard against deadly landslides from the melting slopes above. But many more will remain unprotected and unprepared – until the worst happens, bringing death crashing down from above, suddenly and with no warning. Moreover, this won't be the only danger associated with mountains in the warming world: as later chapters show, just as dangerous will be the likelihood of running out of life's most precious resource – water.

Queensland's frogs boil

No one could accuse the Australian authorities of not taking their responsibility to protect the Queensland Wet Tropics rainforest seriously. Visitors must stay on walking tracks at all times. Fuel for stoves must be brought in, as campfires might disturb the delicate nutrient cycle of the forest. Every tuft of moss, leaf and twig is protected: removing living materials is a criminal offence. Dogs and cats are banned, as are soap, toothpaste and sunscreen, in case these chemicals leach into streams and harm aquatic animals. And you're most certainly *not* allowed to swing from vines in the trees.

There is good reason for this intense conservation focus. Recognised since 1988 as a UNESCO World Heritage Site, seven hundred species of plants are found nowhere else on Earth. The Wet Tropics ecosystem contains many species that are unique relics from ancient rainforests which once grew on the Gondwanan supercontinent 120 million years ago. Many of the same fern species which are still found there today were once grazed by dinosaurs. Amazing carnivorous plants – like pitcher plants and sundews – poke out from the forest floor. Pythons wind around branches, whilst skinks and geckos scurry across rocks and up tree trunks. Thirteen mammal species – including tree kangaroos and ringtail possums – are also unique to the Wet Tropics region. Overall the area is home to a quarter of Australia's frogs, a third of its freshwater fish, and nearly half of its birds – all on a fraction of 1 per cent of the continent.

Yet there is one threat which the Australian authorities are powerless to prevent – and indeed have actively conspired to ignore. It comes not from feral pigs or cane toads, nor even from a thousand swinging, littering, tooth-brushing humans. This threat comes as the climate which sustains the forests – in some areas with a staggering 8 metres of annual rainfall – gradually begins to warm up. It turns out that the Queensland Wet Tropics rainforest is one of the most sensitive areas on the planet to climate change. Just one degree of warming will have devastating impacts on species diversity and habitats.

The reason lies in the unusual topography of the Wet Tropics area. Unlike the Amazon forest, which covers a huge, flat basin until it rises into the eastern slopes of the Andes, the Queensland rainforest comprises hilly terrain – starting from the white sands fringing the ocean to heights of 1,500 metres or more in places. Many of the species which are unique to the area are found only above certain heights: there's a ring-tailed possum which is only found above 800 metres in altitude, and many birds, reptiles and frogs only live at the tops of mountains. As the climate warms, temperature zones rise up the mountainsides, squeezing these species into diminishing islands of habitat – and eventually leaving them with nowhere to live at all. They, like the polar species in the Arctic, will have been literally pushed off the planet.

Dr Steve Williams, of James Cook University's School of Tropical Biology, has been warning for years about the dangers that even small degrees of climate change pose to the Wet Tropics rainforest. Williams – who leads teams of Earthwatch volunteer helpers on his survey trips – has conducted 652 bird surveys, 546 reptile surveys, 342 frog surveys, and at various times set around 50,000 night traps to catch small mammals. Armed with this voluminous wildlife data, he ran a computer model representation of the area under a changing climate and studied the results to see what happened. Even with just a one-degree rise, the results were dramatic. In particular, 63 of the 65 modelled species lost around a third of their core environment. One species of microhylid frog, which instead of having tadpoles in ponds lays its eggs in moist soil, is predicted to go extinct altogether. With higher degrees of warming, rates of biodiversity loss become increasingly dramatic, adding up, in Williams's words, to 'an environmental catastrophe of global significance'.

Nor are animal species the only ones affected. A similar modelling study by David Hilbert of the CSIRO Tropical Forest Research Centre concluded that one degree of warming would reduce the area of highland rainforest by half, wiping out the habitat of many of the rare animal species mentioned above. Rainforests as a whole won't disappear from Queensland as long as the region receives high rainfall, but without these precious throwbacks to an ancient supercontinent, today's world will be immeasurably poorer. Moreover, Australia's national government, which refused for over a decade to take global warming seriously, will have failed in its international duty to protect a UNESCO World Heritage Site.

Just a few miles offshore from the sparkling white sands of the Queensland coast lies another threatened World Heritage Site: the Great Barrier Reef. This is the biggest and most pristine of all the world's coral reefs, a massive subsea wall of coral which is the largest natural feature on Earth, stretching more than 2,300 km along the north-east coast of Australia. One of the most spectacular and diverse ecosystems on the planet, the reef is home to 1,500 species of fish, 359 types of hard coral, 175 bird species and more than 30 types of mammal. It is one of the last refuges of the dugongs (sea cows) and hosts six of the world's seven species of threatened marine turtle.

But the oceans around the Great Barrier Reef are warming – as they are all over the planet – threatening to tip this unique ecosystem into irreversible decline. Coral reefs are actually the external skeletons produced by billions of tiny coral polyps, which secrete calcium carbonate into branches, fans and globes. These constituents in turn combine over thousands of years to form a reef. Each polyp contains algae, tiny plants which live in symbiosis with their animal hosts. Both parties benefit – the coral gets the sugars which the algae produce by photosynthesising light (turning it into energy), whilst the algae derive fertility from the polyp's waste products. But this cosy relationship can only continue in the right aquatic conditions: once the corals' thermal tolerance threshold of 30°C is crossed, the algae are expelled, and the 'bleached' corals will die unless cooler waters return quickly.

Bleaching is undoubtedly a recent phenomenon, observed around the world's oceans only since about 1980. Scientists have drilled deep into reefs and found no evidence that such episodes have happened during past millennia. But as the oceans have warmed due to the human-enhanced greenhouse effect, bleaching episodes have hit the world's coral reefs with increasing – and devastating – regularity. The first mass bleaching event occurred on the Great Barrier Reef in 1998. Since then things have got steadily worse. In 2002 another mass bleaching event occurred – this time 60–95 per cent of all the reefs surveyed across the marine park were bleached to some extent. A small number of reefs, particularly those close to shore where the waters were hottest, suffered almost total wipeout.

As luck would have it, I was on the Great Barrier Reef in the summer of 2002, visiting the University of Queensland's research station on Heron Island. The place was frighteningly efficient – within minutes of disembarking from the Grahamstown catamaran I had learned to tell the difference between a white-capped noddy tern and a muttonbird, and discovered that Heron Island was actually misnamed: the white birds in question are in fact eastern reef egrets. The place was stunning – 'an aviary surrounded by an aquarium', as one of the scientists accurately put it. Buff-banded land rails scampered about like domestic chickens, in and out of the research huts. (Two female students had adopted one and named it Sheryl.) Very soon I spotted the man I had really come to see, striding purposefully around the station with his wetsuit peeled off down to the waist. Ove Hoegh-Guldberg was clearly a man happy in or out of the water. One of his favourite stories was about the time he managed to get his finger clamped in the jaws of a giant clam, and had to rip the animal off the seabed in order not to be drowned by the rising tide – only to be scolded on returning to the beach by a marine park official for damaging a protected species.

Hoegh-Guldberg was the author of a landmark 1999 paper which first drew the world's attention to the threat posed to coral reef survival by bleaching. Having determined the thermal tolerance threshold for corals in different parts of the world, he then applied these to a model of rising sea temperatures during the twenty-first century. The results shocked even him. By the 2020s, he discovered, with less than a degree of global warming, the seas will have heated up so much that the 1998 Barrier Reef mass bleaching event would be a 'normal' year. Given that it takes 30 years or so for a seriously bleached reef to recover, annual bleaching events will devastate the ecosystem – transforming, as Hoegh-Guldberg wrote in the paper, 'Great Barrier Reef communities into ones dominated by other organisms (e.g. seaweeds) rather than reef-building corals'. Other coral reef ecosystems – from the Caribbean to Thailand – would be similarly transformed. With the end of the coral reefs, one of the world's great treasure troves of biodiversity will be destroyed for ever.

It was with this grim scenario in mind that we both went for a snorkel on the afternoon I arrived on Heron Island. Splashing through the shallows, we disturbed a huge shoal of pilchards, which turned en masse and darted off further up the shore. Half a dozen large stingrays flapped lazily further out, where a stronger wind raised enough of a chop to make snorkelling a hazardous experience. Every so often a wave would break over the top of my snorkel, giving me a sudden gulp of salty water. Ove was unfazed, and we trod water for a while as he pointed out affected corals.

'See that bright blue and red? That's actually bleached. It's ironic you get the best colours when it's bleached.' Some of the worst-hit was the branching coral, where whole areas were

36

blanched bone-white. In some places just the tips of the under-water antlers were bleached, whilst in others the entire structure had been hit. But only a minority were the healthy brownish colour indicating the symbiotic algae still at work.

'How likely is it to come back?' I spluttered, swallowing another wave.

'So long as it stays cool from now on, most of it will probably come back,' he replied. 'But some of it won't, and if temperatures rise again soon much of this will probably die.'

Hoegh-Guldberg's work has been complemented by a more recent study which gives a slightly more optimistic forecast. Work in the Caribbean and Indian Ocean by Andrew Baker and colleagues (later published in *Nature*) suggests that corals may be more adaptable – and therefore less vulnerable to outright extinction – than previously thought. The scientists studied coral communities that had been bleached in the 1998 event to see how far they had managed to recover, and were surprised to find that the type of symbiotic algae within the coral had changed to a more heat-tolerant version in all the places they surveyed. With a higher thermal stress threshold than before, damaged reefs may be able to survive future warmer seas without dying out completely, the scientists suggested.

But Ove Hoegh-Guldberg doesn't agree. Even with an increase in heat tolerance with different algae, he points out, ocean temperatures are still set to get too hot for most corals to survive. He and his co-authors used the latest models and reef analysis to again project bleaching frequencies in the decades to come – and their results confirmed the earlier pessimistic analysis. Severe bleaching will occur on most of the world's reefs every 3–5 years by the 2030s, and by the 2050s will strike every two years.

A more recent bleaching event, which struck the Caribbean in 2005, also seems to bear this out. That summer saw sea temperatures in the region reach highs never before measured during the entire 20-year satellite record. These were the same high temperatures, in fact, which made 2005's hurricane season so deadly: Hurricane Katrina hit New Orleans in 2005 after travelling over these same unusually warm areas of ocean. And they were temperatures which would be vanishingly unlikely in an atmosphere without today's loading of greenhouse gases. The effect on Caribbean corals was disastrous. According to surveys carried out by scuba divers, 90 per cent of coral bleached in the British Virgin Islands, 80 per cent in the US Virgin Islands, 85 per cent in the Netherlands Antilles, 66 per cent in Trinidad and Tobago and 52 per cent in the French West Indies. Some reefs may recover in years to come, but model predictions suggest that in this region too bleaching events of this magnitude will hit every other year by the middle of the century.

In any case, very few of the world's reefs are in any state to take on the challenges of climate change. Direct human interference – from sewage, overfishing and agricultural run-off – has already reduced coral reefs across the globe to shadows of their former pristine selves. In total 70 per cent of the world's reefs are now either dead or dying. This is a disaster of an almost unimaginable scale for global biodiversity: second only to rainforests in terms of the vibrancy and diversity of life they nurture, coral reefs world-wide shelter and feed a third of all life in the oceans, including 4,000 types of fish.

Heron Island's reef may be under good management, but the same cannot be said for reefs elsewhere in the Pacific. In the same trip as I visited Ove Hoegh-Guldberg, I also snorkelled along Fiji's so-called Coral Coast, at one of the few gaps I could find between the 5-star hotels and luxury resorts which now blight the entire area. Instead of vibrant-coloured reefs, teeming with parrotfish and groupers, I found piles of rubble – the shattered remains of coral – looming bleakly through a murky ocean. None of the sun-bathers crowding onto the beach seemed to mind, but for me the experience was a depressing reality check. Fiji's Coral Coast is no longer vulnerable to climate change, I was forced to conclude, because it is already dead.

Another hot spot of biodiversity – and yet another World Heritage Site threatened by global warming – is the Cape Floristic Region of South Africa. Covering a huge coastal arc inland from Cape Town, it is home to the greatest concentration of higher plant species in the world outside the tropical rainforests. Its inauspicious rocky soils and arid Mediterranean climate support 9,000 different plants, more than 6,000 of which are found nowhere else on the planet. The most iconic plants in the region are the proteas. The king protea, with its massive sunlike flower head, entirely deserves its designated title as South Africa's national flower. The region is far from pristine, however – lions and rhinoceroses once roamed these hillsides, where now vineyards and rooibos tea plantations encroach on the last wild areas.

According to a team of researchers based at South Africa's National Biodiversity Institute, just small changes in climate could have a devastating impact on the remaining strongholds of the proteas and other endemic species. Using the UK Hadley Centre's model for climate changes in the region by 2020, the scientific team concluded that up to a third of protea species would become threatened or endangered, whilst four would become completely extinct.

In North America too, one degree of climate change could push a threatened species over the brink to extinction – and this one is cute and furry. According to WWF, pikas – small, hamster-like creatures with rounded ears and bushy whiskers – are the first mammal to be endangered by climate change. Pikas live in broken rock on high mountains in the western US and south-western Canada, and are notable not just for being cute and furry, but for their agricultural activities: these small relatives of the common rabbit cut, sun-dry and then store vegetation for winter use in characteristic 'haypiles' on top of rocks. (As a charismatic species, pikas have acquired quite a cult following: check out www.pika works.com for everything from pika music to pika mouse mats.)

However, as the climate warms, pikas – timid beasts, which never stray more than a kilometre from their nests – are set to become increasingly isolated in ever-smaller geographical islands as temperature zones migrate upwards towards the summits. Already local extinctions have been documented at sites in the United States. As the ecologist and pika enthusiast Dr Erik Beever puts it: 'We're witnessing some of the first contemporary examples of global warming apparently contributing to the local extinction of an American mammal at sites across an entire eco-region.'

It has become something of a cliché to talk about the 'canary in the coal mine' when discussing climate impacts on the natural world – but one group of animals more than any other exemplifies this point: the amphibians. With their moist skins and early lives in water, frogs, salamanders and toads are particularly vulnerable to changes in their environment. Indeed, an amphibian – the Costa Rican golden toad – is often cited as the first known case of a global warming extinction.

Once the 'jewel in the crown' of Costa Rica's Monteverde Cloud Forest (to paraphrase the scientist and author Tim Flannery), this Day-Glo orange amphibian was observed in its hundreds back in 1987, gathered around pools in the forest in preparation for mating. But there were already signs of danger: the amphibian expert Marty Crump, who witnessed this last golden toad mating frenzy, also watched the resulting eggs get left behind as the forest pools dried. Only twenty-nine tadpoles lasted out the week, whilst

43,500 eggs were left desiccated and rotting. The following year Crump found only a single, solitary male, and a year later, in 1989, the same male was back once more. That day, 15 May 1989, was the last time anyone saw a golden toad. The species was eventually listed as extinct in 2004. The cause of death seems to have been the general lifting of the mist that nourishes the forest with moist cloud droplets: as the air surrounding the mountains warmed, the cloud base simply rose too high above the forest, allowing the golden toad's spawning pools to dry out.

This memorable animal may be the first, but it is no longer the only amphibian to have gone extinct because of rising temperatures: frog populations have crashed all around the tropics, with more than 100 out of 110 tropical American harlequin frog species having disappeared – even in seemingly pristine forests far away from direct human disturbance. No one knows exactly why: some biologists blame the chytrid fungal pathogen, which is invading new areas and may be causing sudden population crashes. Others blame mystery diseases which are so far undiscovered and unidentified. But experts are largely agreed on one thing: rising temperatures are central to the extinction epidemic, either by helping the new diseases spread, or by stressing amphibian populations and making them more susceptible to die-offs. In this particular murder scene, the weapon may still be in dispute but the overall culprit is clear.

Nowhere, it seems, is safe. One degree of global warming will have severe impacts in some of the world's most unique environments, adding to the biodiversity crisis which is now well under way for reasons unrelated to our changing climate. Pushed out to the margins and isolated in smaller and smaller pockets of natural habitat by ever-expanding zones of human influence, vulnerable wild species will find it impossible to adapt to rapidly changing temperatures by migrating or altering their behaviour.

Whilst coral reefs have a vital role in protecting coastlines from storms and nurturing fisheries, no one can reasonably claim that pikas, proteas and harlequin frogs are essential for global economic prosperity. Their value is intrinsic, not financial. But the world will still be a much poorer place once they're gone.

Hurricane warning in the South Atlantic

With all the headlines about hurricanes hitting the United States, there is one storm that stands out above all others in the way that it caught the scientific community off guard. It wasn't Katrina, which devastated New Orleans and killed over a thousand people. It wasn't Rita, another Category 5 monster which reflooded parts of the city only a month after Katrina struck. Nor was it Hurricane Wilma, which bombed in a single day from being a minor tropical storm into the strongest hurricane ever recorded in the Atlantic basin. No, the storm that really made the forecasters scratch their heads occurred a year earlier, in 2004. And it hit in a part of the world that isn't supposed to experience hurricanes. It was called Catarina, and it struck the coast of Brazil.

Received scientific wisdom holds that hurricanes can only form where sea surface temperatures top 26.5°C. As well as warm oceans, tropical storms also need low 'wind shear': cross-winds at high altitude which can cut the vortex of a developing storm in half. These conditions, as any weatherman can tell you, only occur in the North Atlantic tropics. Not a single South Atlantic hurricane had ever been documented – that is, before March 2004. Indeed, when a strange swirl of clouds began to form off the Brazilian coast on 20 March 2004, local meteorologists couldn't quite believe their eyes. So unheard of was a South Atlantic hurricane that many of them were still refusing to employ the term 'hurricane' when Catarina – complete with 95 mph winds and torrential rain – swept ashore near the town of Torres, damaging 30,000 houses and killing several people. Many of those who suffered, because they also refused to believe that hurricanes were possible in Brazil, had neglected to take shelter as the storm barrelled towards the shore.

In the inevitable meteorological post-mortem, it did indeed look as if the storm was simply a freak, a once-in-a-lifetime experience for those who suffered it. What was strange was that sea temperatures had not been unusually high when it began to gather strength. Instead, what really gave Catarina a boost was a very rare combination of other atmospheric factors which meant that the storm's vortex experienced very little of the deadly wind shear which normally precludes hurricane formation in the South Atlantic. It's a complex picture, but raises an obvious question: will global warming, as well as making the seas warmer and therefore more likely to spark hurricanes, lead more regularly to the conditions which allow tropical cyclones to gather strength in new areas like the South Atlantic?

Two Australia-based meteorologists, Alexandre Bernardes Pezza and Ian Simmonds, addressed this question in their forensic dissection of Catarina which was published in August 2005 by the journal *Geophysical Research Letters*. Their conclusion was tentative, but contained an alarming prospect: it did indeed seem as if the warming atmosphere would favour the conditions which allowed Catarina to form in such an unusual place. 'Therefore,' they wrote, 'there is evidence to suggest that Catarina could be linked to climate change in the southern hemisphere circulation, and other possible future South Atlantic hurricanes could be more likely to occur under global warming conditions.'

Given that if just 0.8°C of global warming so far has already allowed one hurricane to form, a further degree of global warming could make storms in this vulnerable region much more likely in the future. Not only will Brazilians have to batten down the hatches – and perhaps evacuate large areas of their heavily populated coastline – more often, but hurricane-forecasting services will need to be extended to a whole new oceanic basin.

The hurricane season in the following year, 2005, also contained a surprise, which suggests that Brazil is not the only area which will have to keep an eye out for tropical cyclones in our globally warmed future. On 9 October 2005 a new tropical storm appeared about five hundred miles south-east of the Azores, in the East Atlantic, and rapidly gained strength to hurricane status as it drifted past Portugal's Madeira Islands. Hurricane Vince luckily weakened before it made landfall near Huelva in Spain, but set a new record as the first tropical cyclone ever to affect Europe.

Again, conventional wisdom has it that tropical storms can only form over warmer waters thousands of kilometres to the south-west of Iberia. At the time of writing, tropical meteorologists have yet to dissect the unusual combination of factors responsible for Vince, but again the implication is clear: as global warming accelerates, past experience about areas of hurricane formation is not necessarily a reliable guide to the future. Many more hurricane forecasters may be left scratching their heads before they finally admit that not just Brazil but now Europe is already vulnerable to these terrifying storms.

Indeed, there is now evidence for how this might happen: a paper published in July 2007 by Spanish and German climatologists, looking at simulated storms in a computer model, suggests that the whole of the Mediterranean may soon come into the firing line as sea temperatures climb to levels able to spark off genuine tropical cyclones – in a region which has never seen them before. The greatest number of virtual cyclones appeared in the hottest part of the Mediterranean, between Italy and Libya, and once formed, the powerful storms lasted for a week or more.

One computer-generated hurricane formed in the eastern part of the Mediterranean, and then wandered westward all the way to the southern coast of France – much to the amazement of the watching scientists. Another storm formed a tight, symmetrical eye of torrential rain, just as real tropical cyclones do. The idea that once-placid coastlines from Spain to Cyprus could be at risk of landfalling hurricanes in a globally-warmed future has to be one of the most striking projections ever to come from the climate modelling world.

But already real-world evidence is fast emerging that hurricane characteristics are changing as the world's oceans warm up. One of the grandfathers of tropical cyclone physics, Massachusetts Institute of Technology's Kerry Emanuel, recently published a paper in *Nature* that caused its own academic storm. In contrast to the usual view that global warming is still too small a signal to have a measurable impact on tropical cyclones, Emanuel looked again at the data and concluded that storms *were* in fact getting more intense and lasting for longer, due in large part to rising sea temperatures driven by global warming. The storm intensity index hadn't simply gone up by a few percentage points over the last 30 years, either, it had actually doubled – a far greater increase than either theory or modelling had predicted.

Emanuel's data and methods have since been challenged in an academic discussion too technical to analyse here, but it is worth noting that his conclusions are supported by a second piece of work, this time published in *Science* by a team of experts based at the Georgia Institute of Technology in Atlanta. Analysing much of the same storm data – collected by aircraft, satellites and ships over the past three decades – this scientific team identified a large increase in the number and proportion of those hurricanes reaching the strongest categories 4 and 5, despite an overall decrease in the number of cyclones.

Like Emanuel, the team were looking at data from the Pacific as well as the Atlantic Ocean in order to build up a global picture. And like him (though using a different statistical measure), they found a near-doubling in the number of the strongest storms between 1970 and 2004. The Georgia team concluded that the upswing in Category 4 and 5 hurricanes is unlikely to be the result of natural climate cycles, but instead is probably connected to rising temperatures in the tropical oceans.

A year later, after the record-breaking 2005 Atlantic hurricane season – which left 1,000 people dead, 1 million homeless, and caused $200 billion of damage – two leading climatologists tried to settle the argument about whether or not global warming had contributed to the run of disastrous storms. Noting that the warm sea temperatures measured that year – the highest ever – undoubtedly contributed to the ferocity of Katrina, Wilma, Rita and the other 2005 storms, Kevin Trenberth and Dennis Shea used complex maths to figure out how much of the Atlantic warming signal was due to global warming, and how much to natural cycles. Their conclusion should be a wake-up call to us all: at least half of the extra warmth had come from human-caused global warming. As so many people suspected at the time, Katrina was only partly a natural disaster.

Sinking atolls

I hate to put this so bluntly, but in all probability nothing can save the Pacific island of Tuvalu. Like a slowly boiling kettle, the oceanic system has very long response time to changing conditions, and the seas will go on slowly rising for centuries even if all greenhouse gas emissions stopped tomorrow. With Tuvalu already experiencing regular flooding events due to past sea level rise – as I documented in *High Tide* – this extra rise in the world's oceans will sound the death knell for this fascinating and lively island society.

Tuvalu, with only 9,000 inhabitants, is actually one of the smallest of the five atoll nations which will shortly cease to exist. The others are Tuvalu's sister atoll group Kiribati, with 78,000 population, the Marshall Islands, with 58,000 people, tiny Tokelau (2,000 people; a dependent territory of New Zealand) and the Maldives, the largest and most densely populated of all the island groups, with 269,000 inhabitants. Together with people displaced from coastal areas of other non-atoll islands, this already totals about half a million people who – suddenly divorced from their cultures and their origins – will need to find new homes. New Zealand has hesitantly offered to take a small number of Tuvaluans, but no other nations have yet stepped up to offer themselves as places of refuge, least of all those rich countries who have done most to cause the problem in the first place.

Unless spurred on by a major hurricane or storm surge, the end for atoll countries will not be rapid or cathartically dramatic. Instead it will be death by a thousand cuts, an incremental diminishment of each nation's ability to support itself, as young people lose confidence in the future and old people sink back into comforting dreams of the past. Each bit of beach lost, each vegetable garden invaded by salt water, each undercut coconut tree which topples into the waves will add to the inevitable toll. Decades before the last bit of coral disappears under the sea, community services will decline, children will emigrate, schools will close, and the fabric of a nation will begin to unravel.

Bear in mind that as future chapters of this book unfold, unacknowledged and mostly forgotten, atoll nations will be submerging – bit by bit.

2°

2

TWO DEGREES

China's thirsty cities

Take a train from Hohhot to Lanzhou in northern China, and you pass through a strange area of heavily eroded badlands, where steep gullies and cliffs crowd in around the railway track as it weaves its way through a narrow river gorge. At many points caves have been hacked out of the cliffs – their history is murky, but perhaps they were used by vagrants, or people expelled from the cities, or even by Communist dissidents exhorting the peasantry to rise up against the Nationalists during the 1930s. A more prosaic explanation is that they were carved out by railway construction workers as they laboured to lay track through cold, windy and inhospitable territory.

These badlands are the edge of China's loess plateau, a gigantic area of compacted dust many hundreds of metres deep, deposited over thousands of years by dust storms and strong winds roaring down from the Gobi Desert of Mongolia. This dry plateau may not be much good for agriculture, grazing or anything else (bar digging caves) but it is a treasure trove for palaeoclimatologists, who use its finely preserved layers of dust and sand to reconstruct the fluctuations of ancient climates across the whole region of northern China.

51

It was with this purpose in mind that a team of Chinese scientists based in Lanzhou trekked out to four sites on the loess plateau in 1999, drilling more than 30 metres down into the compacted soil before carefully extracting their sections and carting them back to the lab. Near the base of each section was the target of the research: a layer of prehistoric soil – 'palaeosol' in the jargon – dating from the Eemian interglacial, the previous warm period before the start of the last ice age. The weather records preserved in this inauspicious red-brown layer would prove to hold clues not just about the past, but also about the future.

Like Africa and the Indian subcontinent, northern China is subject to an annual monsoon cycle. In summer, moist air blows in from the ocean, bringing heavy rainfall to the south. In winter, however, the pattern reverses and strong winds sweep down from the north, bringing dust and freezing temperatures. The Lanzhou scientists, using complex techniques to measure particle sizes and magnetic data from the palaeosol section, were able to draw conclusions from their sample about changing monsoon strength 129,000 years ago, as the Eemian climate gradually warmed up. Because of how long it takes the oceans to absorb heat, it appeared, the dry winter monsoon responded much more rapidly than the summer one to the changing conditions. The result was a period of drought and continental-scale dust storms, before the summer monsoon pushed far enough inland to bring significant rainfall to the loess plateau.

So could such a mechanism plausibly repeat in the two-degrees-warmer world? Studies of Pacific sea floor sediments suggest that the height of the Eemian interglacial saw global temperatures about 1°C higher than today's, making this period a potentially useful analogue for a warmer climate in the future – particularly as regional temperatures in a large continent like Asia would in turn have been a degree or so higher than the global average. And if it did take China's monsoon climate longer to make the transition from cool/dry to warm/wet 129,000 years ago, as some scientists believe, this does suggest a possible cause for the droughts and rising temperatures that have struck northern China in recent years. So whilst southern China can expect more flooding as the climate warms, the oceanic time lag means that it may take much longer for the rain-bearing summer monsoon to reach the drought-stricken north. With China split between two extremes, agriculture will inevitably suffer, and water-stressed cities like Beijing and Tianjin will continue to experience shortages – particularly as economic growth spirals upwards and underground aquifers are pumped dry. The Chinese government has begun construction of a massive water transfer project, which aims to take billions of cubic metres of water from the Yangtze River in the south to the thirsty cities in the north. However, even this mega-project – the largest ever constructed on the planet – will (if it works, which many doubt) have difficulty keeping the taps running. With a chronic shortage of water, China will not just struggle to develop a more affluent lifestyle, it will struggle to feed itself too.

Acidic oceans

Greenhouse gases released over the last hundred years or so have not only changed the climate; they have also begun to alter conditions in the largest planetary habitat of all – the oceans. At least half the carbon dioxide released every time you or I jump on a plane or turn up the air-conditioning ends up in the oceans. This may seem like a good place for nature to dump it, but ocean chemistry is a complex and delicate thing. The oceans are naturally slightly alkali, allowing many animals and plants which inhabit the seas to build calcium carbonate shells.

However, carbon dioxide dissolves in water to form carbonic acid, the same weak acid that gives you a fizzy kick every time you swallow a mouthful of carbonated water. That's great for a glass of Perrier, but not so good if it's beginning to happen on a gigantic scale to the whole of the global ocean. And it is indeed beginning: humans have already managed to reduce the alkalinity of the seas by 0.1 pH units. As Professor Ken Caldeira of the Carnegie Institution's global ecology department says: 'The current rate of carbon dioxide input is nearly 50 times higher than normal. In less than 100 years, the pH of the oceans could drop by as much as half a unit from its natural 8.2 to about 7.7.' This may not sound like much, but this half point on the pH scale represents a fivefold increase in acidity. And because the oceans circulate only very slowly, even if atmospheric carbon dioxide levels are eventually stabilised – perhaps because humanity wakes up to their warming effect – these changes in ocean chemistry will persist for thousands of years.

This fast-moving area of scientific research was the subject of a major report by the Royal Society in June 2005, which identified some of the main concerns that are increasingly keeping marine biologists awake at night. First and foremost is the possibility that even with relatively low future emissions during this century (equating to two degrees or less of temperature warming), large areas of the Southern Oceans and part of the Pacific will become effectively toxic to organisms with calcium carbonate shells after about 2050. With higher emissions, indeed, most of the entire global ocean will become eventually too acidic to support calcareous marine life.

The most important life-forms to be affected are those that form the bedrock of the oceanic food chain: plankton. Although individually tiny (only a few thousandths of a millimetre across), photosynthesising plankton like coccolithophores are perhaps the most important plant resource on Earth. They comprise at least half the biosphere's entire primary production – that's equivalent to all the land plants put together – often forming blooms so extensive that they stain the ocean surface green and can easily be photographed from space. The places where phytoplankton thrive are the breadbaskets of the global oceans: all higher species from mackerel to humpbacked whales ultimately depend on them. Yet coccolithophores have a calcium carbonate structure, and this makes them especially vulnerable to ocean acidification. When scientists simulated the oceans of the future by pumping artificially high levels of dissolved CO_2 into sections of a Norwegian fjord, they watched in dismay as coccolithophore structures first corroded, and then began to disintegrate altogether.

Acidification will also directly affect other ocean creatures. Crabs and sea urchins need their shells to survive, whilst fish gills are extremely sensitive to ocean chemistry – just as our lungs are to the air. Mussels and oysters, vitally important both as economic resources and as part of coastal ecosystems worldwide, will lose their ability to build strong shells by the end of the century – and will dissolve altogether if atmospheric CO_2 levels ever reach as high as 1,800 ppm (parts per million: this ppm measure means, very simply, that for every million litres of air there are 1,800 litres of carbon dioxide). Tropical corals, already badly hit by bleaching, will more and more be corroded by this increasing oceanic acid. Walk out to sea on a reef in 2090, and it may crumble beneath your feet. Ships, rather than being torn apart when they strike rocky coral, may find themselves ploughing through weakened reefs like sponge. Indeed, it's difficult to overstate just how dangerous an experiment we are now conducting with the world's oceans. As one marine biologist says: 'We're taking a huge risk. Chemical ocean conditions 100 years from now will probably have no equivalent in the geological past, and key organisms may have no mechanisms to adapt to the change.'

Phytoplankton are also crucial to the global carbon cycle. Collectively they are the largest producer of calcium carbonate on Earth, removing billions of tonnes of carbon from circulation as their limestone shells rain down onto the ocean floor. There's nothing new about this process: the chalk in the cliffs and downs of southern England originally formed as the limey sludge from countless billions of dead coccolithophores back in the Cretaceous era. But as the oceans turn more and more acidic, this crucial component of the planetary carbon cycle could slowly grind to a halt. With fewer plankton to fix and remove it, more carbon will remain in the oceans and atmosphere, worsening the problem still further.

Phytoplankton are also hit directly by rising temperatures, because warmer waters on the surface of the ocean shut off the supply of upwelling nutrients that these tiny plants need to grow. As with acidification, changes are already detectable today: in 2006 scientists reported a decline in plankton productivity of 190 megatonnes a year as a result of the current warming trend. Together these two factors, warming and acidification, represent a devastating double blow to ocean productivity. As Katherine Richardson, professor of biological oceanography at Aarhus University in Denmark, says: 'These marine creatures do humanity a great service by absorbing half the carbon dioxide we create. If we wipe them out, that process will stop. We are altering the entire chemistry of the oceans without any idea of the consequences.'

Wiping out phytoplankton by acidifying the oceans is rather like spraying weedkiller over most of the world's land vegetation, from rainforests to prairies to Arctic tundra, and will have equally disastrous effects. Just as deserts will spread on land as global warming accelerates, so marine deserts will spread in the oceans as warming and acidification take their unstoppable toll.

The mercury rises in Europe

Under normal circumstances, the human body is good at dealing with excess heat. Capillaries under the skin flush with blood, allowing the extra warmth to radiate into the air. Sweat glands pump out moisture, disposing of heat through evaporation. Heat can even be lost through panting, and the heart works overtime. During exercise, the normal body temperature of 37 degrees Celsius can rise to 38 or 39°C with no ill effects.

But 2003 was not a normal summer, and the heatwave experienced in Europe during the three months of June, July and August did not produce normal circumstances. In Switzerland the mercury climbed above 30°C as early as 4 June, rising to a maximum of 41.1°C in the south-east of the country on 2 August – the sort of searing temperature more often associated with the Arabian Desert than temperate central Europe. Across the continent, records tumbled: in Britain temperatures reached 100° Fahrenheit for the first time. Beaches were packed as holidaymakers enjoyed the summer heat, but in big cities like Paris, a hidden disaster was unfolding.

The first symptoms of heat stress may be minor. An affected individual will feel slightly nauseous and dizzy, and perhaps get irritable with those around. This needn't yet be an emergency: an hour or so lying down in a cooler area, sipping water, will cure early heat exhaustion with no longer-lasting symptoms. But in Paris, August 2003 there were no cooler areas, especially for elderly people cooped up in their airless apartments. It wasn't so much the high temperatures of the day, but the fact that things didn't cool down enough at night to give the body time to recover. The effects were cumulative, and the most dangerous – and often fatal – form of heat stress then became much more likely: hyperthermia or heatstroke.

Once human body temperature reaches 41°C (104°F) its thermoregulatory system begins to break down. Sweating ceases, and breathing becomes shallow and rapid. The pulse quickens, and the victim may rapidly lapse into a coma. Unless drastic measures are taken to reduce the body's core temperature, the brain is starved of oxygen and vital organs begin to fail. Death will be only minutes away unless the emergency services can quickly get the victim into intensive care.

These emergency services failed to save over 10,000 Parisian heatstroke victims in the summer of 2003. Mortuaries quickly ran out of space as hundreds of dead bodies, mainly of elderly and marginalised people, were brought in each night. The crisis caused a political furore as people accused politicians and municipal administrators of being more concerned with their long August holidays than with saving lives in the capital. Estimates vary, but across Europe as a whole, between 22,000 and 35,000 people are thought to have died.

The heatwave and drought also devastated the agricultural sector: crop losses totalled around $12 billion, whilst forest fires in Portugal caused another $1.5 billion of damage. Major rivers such as the Po in Italy, the Rhine in Germany and the Loire in France ran at record low levels, grounding barge traffic and causing water shortages for irrigation and hydroelectric production. Toxic algal blooms proliferated in the denuded rivers and lakes. Melt rates on mountain glaciers in the Alps were double the previous record set in 1998, and some glaciers lost 10 per cent of their entire mass during the heat of that one summer. Meanwhile – as described in chapter 1 – melting permafrost caused rockfalls in mountain areas like the Matterhorn.

It wasn't long before questions were asked about global warming's possible contribution to the disaster. Meteorologists who investigated past hot spells found that the 2003 heatwave was off the statistical scale – a one-in-several-thousand-year event. According to an analysis by UK-based climatologists, twentieth-century global warming has already doubled the risk of such a heatwave occurring. Right across Europe, according to research published in 2007, the frequency of extremely hot days has tripled over the last century, and the length of heatwaves on the Continent has doubled. The conclusion is stark: the 2003 summer hot spell was not a natural disaster.

The intensity of the heatwave also tells us something about the future. Averaged across the whole continent, temperatures were 2.3°C above the norm. So does that mean that in the two-degree world, summers like 2003 will be annual events? It seems so: in the UK-based study mentioned above, scientists used the Met Office's Hadley Centre computer model to project future climate change with increasing greenhouse gas emissions, and concluded that by the 2040s – when temperatures globally in their model are still below two degrees – more than half the summers will actually be warmer than 2003.

That means that extreme summers in 2040 will be much hotter than 2003 – and the death toll will rise in consequence – perhaps reaching the hundreds of thousands. Elderly people may have to be evacuated for months at a time to air-conditioned shelters, and outside movement during the hottest part of the day will become increasingly dangerous. Temperatures may soar to highs commonly experienced today only in North Africa, as rivers and lakes dry up and vegetation withers across the entire continent. Crops which require summer rainfall will bake in the fields, and forests which are more accustomed to cooler climes will die off and burn. As a result, catastrophic wildfires may penetrate north into new areas, torching broadleaved forests from Germany to Estonia.

Here again the summer of 2003 gives us a glimpse of things to come. Europe-wide monitoring systems showed a 30 per cent drop in plant growth across the continent, as photosynthesis began to shut down in response to the twin stresses of high temperatures and crippling drought. From the deciduous beech forests of northern Europe to the evergreen pines and oaks of the Mediterranean rim, plant growth across the whole landmass slowed and then stopped. Instead of absorbing carbon dioxide from the air, the stressed plants instead began to emit it; around half a billion tonnes of carbon was added to the atmosphere from European plants, equivalent to a twelfth of total global emissions from fossil fuels. This is a positive feedback of critical importance, because it suggests that as temperatures rise – particularly during extreme heatwave events – carbon emissions from forests and soils will also rise, giving a further boost to global warming. And if these land-based emissions are sustained over long time periods and large areas of the Earth's surface, global warming could begin to spiral out of control, as the next chapter shows.

We may have come dangerously close to that point during the 1998–2002 mid-latitudinal drought in the northern hemisphere, which left plants withering through regions as far afield as the western US, southern Europe and eastern Asia. One study showed that carbon emissions which would normally have been taken up by plants instead accumulated in the atmosphere, explaining the abnormally large jumps in the atmospheric CO_2 concentration in following years. (Jumps which caused jitters among many climate change watchers about whether runaway positive feedbacks might have already begun.) Over a billion tonnes of extra carbon poured out of plants and soils in response to the drought and heat.

At the time of writing, the heatwave of 2003 has already begun to fade in people's memories, and the 'normal' summers of the following two years will have begun to soak up some of the extra carbon that entered the atmosphere during that deadly hot spell.

But we forget at our peril. The summer of 2003 was a 'natural experiment' whose conclusions should be taken very seriously. This wasn't just some output from a computer model, whose assumptions and projections can be legitimately challenged. It actually happened. Moreover, the near-repeat of the 2003 heatwave in the summer of 2006 suggests that if anything the models are underestimating the likely frequency and severity of future heatwaves.

We have been warned.

Mediterranean sunburn

Perhaps the most striking images from 2003's hot summer came from Portugal, where gigantic forest fires swept through the tinder-dry landscape, destroying orchards, torching houses and killing eighteen people. In total an area almost the size of Luxembourg was devastated. The conflagrations were so huge that they cast palls of smoke right over the North Atlantic, with both fires and smoke easily visible from space. The fires must have been particularly shocking for tourists, many of whom flock to southern Portugal from northern Europe – more in search of the sun than several days of smoke inhalation.

However, one study shows that such wildfires are going to be an increasingly common sight for holidaymakers to southern Europe and the Mediterranean. Climate change simulations show the region getting drier and hotter as the subtropical arid belt moves northward from the Sahara. In the two-degree world, two to six weeks of additional fire risk can be expected in all countries around the Mediterranean rim, with the worst-hit regions being inland from the coast where the temperatures are highest. In North Africa and the Middle East virtually the whole year will be classified as 'fire risk'.

These fires will be driven on by scorchingly hot temperatures.

The number of days when the mercury climbs over 30°C is expected to increase by five to six weeks in inland Spain, southern France, Turkey, northern Africa and the Balkans. The number of 'tropical nights', when temperatures don't cool off past 20°C, will increase by a month, and the entire region can expect an additional four weeks of summer. A doubling of what the study calls 'extremely hot days' is also projected, whilst land areas around the Mediterranean can expect three to five additional weeks of 'heatwaves' (defined as days with temperatures over 35°C). Islands such as Sardinia and Cyprus only tend to escape the worst because of the cooling influence of the sea.

The high temperatures will be aggravated by drought, with areas in the southern Mediterranean projected to lose around a fifth of their rainfall. Spain and Turkey will also be badly affected, whilst northern areas on average see a 10 per cent decline in rainfall and a corresponding two- to three-week increase in the number of dry days. Up to a month extra of drought can be expected in southern France, Italy, Portugal and north-west Spain. The seasonality of rainfall will also change, playing havoc with agricultural practices: in southern France and Spain, for example, the dry season is projected to begin three weeks earlier and end two weeks earlier.

Air-conditioning may not always be an option: with peak power demand occurring during the driest part of the year when reservoir levels are already low, hydroelectric power outages could lead to blackouts during the worst heatwaves. Tourists – especially the elderly – will need to stay away because of the danger of heatstroke, whilst Mediterranean locals might actually prefer to spend summers far away in northern Europe in search of cooler temperatures. Lifestyles will have to change, with people perhaps adopting more Middle Eastern or North African living routines to cope with the heat.

Water shortages will become a perennial problem around the whole Mediterranean basin, particularly as some of the most arid coastal areas of Spain and Italy are also some of the most densely populated. Rich Germans and Britons thinking of retiring to Spain might be well advised to stay put. With searing heat and little fresh water to cool things off, perhaps the lure of the sun won't be so strong after all. The mass movement in recent decades of people from northern Europe to the Mediterranean is likely in the two-degree world to begin to reverse, switching eventually into a mass scramble to abandon barely habitable temperature zones – as Saharan heatwaves sweep across the Med.

The coral and the ice cap

Back in 1998, three Canadian geologists took a trip to the Cayman Islands. They were not there to sunbathe or launder money (two activities for which the islands are justly famous) but to investigate a strange raised limestone platform in the Rogers Wreck area of Grand Cayman island. The platform – known to geologists as the Ironshore Formation – is about 20 metres thick, and includes layers of ancient coral hundreds of thousands of years old. The formation sparked the scientists' interest because if they could date the coral accurately, its height above sea level today would help them solve a mystery about how sea levels had changed in the past.

Tropical coral reefs form in shallow seas, so if old coral is now above sea level, only two explanations are possible: the land has risen, or the sea level has fallen. After meticulous investigation, the three scientists – Jennifer Vezina, Brian Jones and Derek Ford of the University of Alberta's Earth and Atmospheric Sciences department – ruled out land uplift and concluded that sea levels during the previous Eemian interglacial period were many metres higher than they are now.

The Canadian scientists' conclusion chimed with other studies from around the world, which have also suggested that sea levels were 5–6 metres above present during the Eemian, 125,000 years ago. Given that global temperatures were then about 1°C higher than now (though slightly higher in the Arctic, thanks to the polar amplification effect), this in turn raised another question: where had all the extra water come from?

First to come under suspicion was the West Antarctic Ice Sheet. Glaciologists had long suspected that it might be sensitive to small changes in temperature, and in total it contains enough ice to raise global sea levels by 5 metres. Indeed, as early as 1978 a paper in *Nature* warned that the ice sheet posed 'a threat of disaster' – a warning which is even more pressing today, as chapter 4 reveals. But attempts to model ice sheet collapse had proven inconclusive, and in 2000 an entirely different contributor to sea level rise was proposed: Greenland.

The Greenland ice cap contains enough water in its 3-kilometre-thick bulk to raise global sea levels by a full 7 metres, and when scientists investigated cores drilled from the summit of the ice sheet they reached a surprising conclusion. Greenland had indeed shrunk significantly during the Eemian – so much so in fact that most of the southern and western part of the landmass had been completely free of ice for thousands of years. Indeed, evidence has recently emerged that Greenland was once forested in regions that are now under two kilometres of ice – although this may have been in an earlier (and slightly warmer) interglacial than the Eemian. With a lower summit, steeper sides and a drastically reduced extent, the Eemian ice sheet would have contributed, the scientists concluded, between 4 and 5.5 metres to higher global sea levels at the time. This, together with smaller contributions from Antarctica and other glaciers, plus some thermal expansion of seawater, would seem to explain the high sea levels.

The study raised a few academic eyebrows at the time, but its implications didn't really begin to sink in until several years later. In retrospect, this is perhaps surprising: it contained clear evidence that a climate only a degree or so warmer than today could melt enough Greenland ice to drown coastal cities around the globe, cities that are home to tens of millions of people. Nor was it just a one-off: more recent work confirms that Greenland's contribution to the higher sea levels of the Eemian was indeed somewhere between 2 and 5 metres.

The 2001 report by the Intergovernmental Panel on Climate Change (IPCC) did conclude that higher temperatures would eventually melt the Greenland ice sheet – but only over centuries to millennia, and very little contribution from Greenland was factored into the twenty-first-century sea level rise projections of between 9 and 88 cm. As warnings go, it wasn't a terribly urgent one: most people have trouble caring about what happens 100 years hence, let alone bothering about whether their distant descendants in the year 3000 might be getting their feet wet.

One man begged to differ, and he wasn't some sandal-wearing greenie who could be easily dismissed. The new warning came from James Hansen, the NASA scientist whose testimony to Congress back in the hot summer of 1988 did so much to put global warming on the international agenda for the first time. Hansen penned a characteristically straightforward article entitled 'Can we defuse the global warming time-bomb?', later published in *Scientific American*, which asked the key question: 'How fast will ice sheets respond to global warming?' The article was critical of the IPCC's assurances that ice sheet melting would be gradual even in a rapidly warming world, words which Hansen felt had downplayed the urgency of our situation.

Hansen noted instead that a global temperature rise over 1°C could destabilise the polar ice sheets enough to give rises in sea levels far greater than the modest 50 centimetres or so by 2100 that is seen as most likely by the IPCC. At the end of the last ice age, for example, global sea levels shot up by a metre every twenty years for a period of four centuries, drowning tropical coral reefs in Hawaii and submerging low-lying coasts. This dramatic flood, termed 'Meltwater Pulse 1a' by scientists, occurred 14,000 years ago as the giant ice sheets of the last glacial age finally crumbled and gave way to the warmer Holocene.

What has happened before can happen again, argued Hansen, especially given today's enormous atmospheric loading of greenhouse gases, whose climatic impact far outweighs the tiny orbital changes which govern the ice-age-to-interglacial transitions. Just as they were in the past, ice sheet changes in the future could be – to use Hansen's phrase – 'explosively rapid'.

Hansen had little support, however, until the following year, when a European modelling team put an actual figure on Greenland's critical melt threshold: 2.7°C. This, moreover, wasn't a figure for *global* warming but instead for *regional* warming. Because the Arctic heats up faster than the globe as a whole, this tipping point will be crossed sooner in Greenland than the global average: because of polar amplification, reported a second scientific team, Greenland warms at 2.2 times the global rate. Divide one figure by the other, and the result should ring alarm bells in coastal cities across the world: Greenland will tip into irreversible melt once global temperatures rise past a mere 1.2°C.

That's the bad news. The good news is that according to this study the Greenland ice sheet will contract only slowly, over millennia, to a smaller inland form. With higher levels of warming (up to 8°C regionally, for example, if greenhouse gas emissions continue to rise unabated) most of the ice sheet will disappear over the next 1,000 years, still giving humanity plenty of time to prepare for the full 7 metres of inundation – even though lower-lying areas would go under much sooner.

In addition, some of the melting will be offset by increased snowfall, leading to thicker ice in the centre. This is another result of rising temperatures, as a warmer atmosphere can hold more water vapour. Much of Antarctica and the interior of Greenland is classed as 'polar desert' because it is simply too cold for snow to fall in significant amounts. Already evidence suggests that areas of the Greenland ice sheet above the 1,500-metre contour line are accumulating snow and new ice (at 6 cm a year according to one study). It has even been suggested that a thicker Greenland ice sheet could offset rising sea levels.

But real-world evidence runs counter to these optimistic scenarios, suggesting that James Hansen may be right after all. The models on which the predictions of Greenland ice melt are based operate by estimating the difference between water loss from future melting and ice accumulation from future snowfall. There is much more to ice sheet dynamics than just melting and snow-fall, however. Vast quantities of ice are constantly flowing away from Greenland's centre in gigantic glaciers, which surge through fjords and discharge icebergs into the sea. These glaciers can have a rapid effect on the stability of the ice sheet, yet they aren't properly accounted for in the models. 'Current models treat the ice sheet like it's just an ice cube sitting up there melting, and we're finding out it's not that simple,' says Ian Howat, an expert on Greenland's glaciers.

In particular, as melting proceeds on the surface, whole rivers plunge down through icy sink-holes called moulins onto the bedrock beneath the ice sheet. This meltwater then acts as a lubri-cant underneath the ice, speeding up the glaciers as they proceed towards the sea. As one glaciologist told *Nature*: 'Along the coasts, all the glaciers are thinning like mad, and they're also flowing faster than they ought to. Changes initiated in coastal regions will propagate inland very quickly.' When Byron Parizek and Richard Alley, two glaciologists at Penn State University in the US, had a first stab at including meltwater lubrication in an ice sheet model, they found that this did indeed lead to a thinner ice cap over Greenland, and a greater contribution to sea level rise.

Greenland's glaciers are also changing much quicker than anyone expected. The largest outflow glacier on the whole landmass, Jacobshavn Isbrae in the south-west, is so huge that it alone has a measurable impact on global sea levels, accounting for 4 per cent of the twentieth-century rise. Not only has the gigantic river of ice thinned by a phenomenal 15 metres annually since 1997 (that's about four office block storeys every year), but its flow rate more than doubled between 1997 and 2003, suggesting that an increased quantity of Greenland's ice is being sucked out into the sea. As if to emphasise the abnormal shift, Jacobshavn Isbrae's floating ice shelf has now suffered almost complete disintegration, spawning an armada of icebergs along the coast.

On the eastern side of the ice cap, a second glacier has also suffered dramatic changes. A US-based research team led by Ian Howat studied satellite photos of the Helheim Glacier's behaviour between 2000 and 2005, and was stunned to discover that not only had the ice flow speeded up, but the glacier had thinned by over 40 metres and retreated several kilometres up its fjord. About half of the thinning is due to increased surface melt: in recent years ever-wider areas of Greenland have been rising above freezing, and thousands of blue meltwater pools now pepper the ice surface in the summer. But the remainder is due to the speeding-up of glacial flow rates, to the distinctly unglacial pace of 11 kilometres per year.

This faster flow draws more ice down the valley, thinning the glacier just as a rubber band gets thinner when you stretch it. In the process, now being repeated in glaciers right around the ice cap, billions of tonnes more ice are dumped in the North Atlantic, raising sea levels still further. According to Howat, the thinning has reached a 'critical point' which has begun 'drastically changing the glacier's dynamics'. His conclusion is devastating: 'If other glaciers in Greenland are responding like Helheim, it could easily cut in half the time it will take to destroy the Greenland ice sheet.'

Other scientists concur. Speaking at the Fall Meeting of the American Geophysical Union in December 2005, the University of Maine's Dr Gordon Hamilton also noted 'very dramatic changes' on eastern Greenland's Kangerdlugssuaq Glacier. In just one year, between April 2004 and April 2005, this enormous glacier both doubled in speed and simultaneously retreated by 4 kilometres. If other large glaciers begin to go the same way as Helheim and Kangerdlugssuaq – which have now doubled the rate at which they together dump ice in the ocean from 50 to 100 cubic kilometres a year – it could 'pull the plug' on Greenland, Dr Hamilton warned.

The newest evidence suggests that all is not yet lost however. In March 2007 Ian Howat and colleagues reported in *Science* magazine that results from their latest survey work might be more reassuring. Although both the Helheim and Kangerdlugssuaq glaciers had indeed doubled their rate of mass loss in 2004, as previously reported, two years later – by 2006 – they had returned to something more like normality. However, the glaciologists Martin Truffer and Mark Fahnestock, writing in *Science* magazine in March 2007, are at pains to point out that this most recent change 'does not mean that [the glaciers] have stabilised'. Instead, 'the question remains whether changes in the past five years have left the system as a whole more vulnerable'. Clearly these giant rivers of ice are complex beasts that scientists are still struggling to understand.

But whatever the behaviour of individual outlet glaciers, wider-scale satellite studies of Greenland's entire ice cap do suggest that major changes are afoot. Scientists working on the GRACE satellite programme (Gravity Recovery and Climate Experiment) reported in November 2006 that big losses are now under way. Whilst the ice sheet was probably in balance for much of the 1990s, between 2003 and 2005 it shed about 100 billion tonnes a year of ice, enough to raise global sea levels by 0.3 mm a year.

Jim Hansen, for one, is continuing his battle to get the world to wake up to the threat of melting ice caps, and attempts by the Bush administration and Hansen's bosses in NASA to silence him have been given characteristically short shrift. (When NASA public relations staff ordered him to stop giving lectures or talking to journalists without clearing material with them first, Hansen went straight to the press with the story, sparking embarrassing 'NASA censorship' headlines across the US media.) His publications – even in weighty scientific journals – are now increasingly peppered with strong words such as 'dangerous' and 'cataclysm', ignoring the usual convention that scientists must muzzle themselves with emotionless jargon. One Hansen paper published (with five leading co-authors) in May 2007 warns bluntly in the abstract that 'recent greenhouse gas emissions place the Earth perilously close to dramatic climate change that could run out of our control, with great dangers for humans and other creatures' – a clear statement of fact, but one which must have raised disapproving eyebrows throughout the halls of academe.

However, Hansen's contention that the world's ice sheets could collapse much more rapidly than the IPCC suggests does have a solid base in physics. In order to explain the rapid real-world fragmentation of ice sheets at the end of the last ice age, Hansen outlines a process called the 'albedo-flip' – something which, if repeated today, could destroy the remaining ice sheets much faster than conventional projections suggest. This albedo-flip is worryingly simple: as snow and ice melt they become wet, making the surface darker and therefore more able to absorb sunlight. This raises temperatures further, sparking wider melting, in a classic positive feedback. The albedo-flip is why, Hansen suggests, ice sheet disintegration can be 'explosively rapid' rather than a more stately process taking millennia to play out. And given that large areas of Greenland and west Antarctica are already bathed in summer meltwater, Hansen suggests that this 'trigger mechanism' of darker, wet snow is already engaged today.

So how fast might sea levels rise? The IPCC's 2007 report suggests only 18 to 59 cm – reassuring figures to those who live close to the coast. However, it also introduces a caveat, admitting that uncertainties about ice sheet response times could make this figure higher. But it doesn't say how much higher, and no one else in the scientific community has ventured to offer an estimate – except, once again, James Hansen whose warning about global warming as long ago as 1988 suggests remarkable foresight. In a paper entitled 'Scientific reticence and sea level rise' and published in 2007 in the online free journal *Environmental Research Letters*, Hansen takes his colleagues to task for staying 'within a comfort zone' and refusing to say anything which may prove to 'be slightly wrong'. Instead, he argues, 'There is enough information now, in my opinion, to make it a near certainty that IPCC business-as-usual climate forcing scenarios would lead to a disastrous multi-metre sea level rise on the century time scale.' If the ice sheet melt rate doubles each decade, a serious possibility, the resulting sea level rise will total 5 metres by 2100, he points out.

So perhaps James Hansen's warnings should be taken more seriously, particularly his concern that melt rates and sea level rise could speed up dramatically over the century to come. An early warning is already out there: sea levels are currently rising at 3.3

mm a year – much faster than the 2.2 mm projected by the IPCC's 2001 report. If, as Hansen suggests is likely, melt rates as rapid as those at the end of the last ice age begin to happen again this century, the whole Greenland ice sheet could disappear within 140 years. The geography of the world's coastlines would then look radically different. Miami would disappear entirely, as would most of Manhattan. Central London would be flooded. Bangkok, Bombay and Shanghai would also lose most of their area. In all, half of humanity would have to move to higher ground, leaving landscapes, buildings and monuments that have been central to civilisation for over a thousand years to be gradually consumed by the sea.

Last stand of the polar bear

Not everyone sees the transformation of the Arctic as a bad thing. Even as water sluices off the melting Greenland ice sheet, climate change at the top of the world could be making some people very rich. Pat Broe hopes to be one of them. An American entrepreneur, in 1997 Broe bought the run-down north Canadian port of Churchill for the princely sum of $7. For the fewer than a thousand residents who live in this drab-looking town, life has been tough – even in a place which has rebranded itself as the 'Polar Bear Capital of the World'.

But according to Broe, boom times lie just around the corner. As the polar ice melts, humble little Churchill could become an important hub in lucrative shipping routes opening up between Asia, Europe and North America across waters that used to be permanently frozen. Which is just as well, because by the time these shipping routes open, the Polar Bear Capital of the World will be looking for another raison d'être – for the simple reason that as the sea ice disappears, so will the polar bears.

One rather unamusing irony of global warming is that the retreat of the northern polar ice cap is sparking a new petroleum gold rush, bringing further fossil fuels onto world markets which – when burned – will inevitably make the climate change problem worse. According to some estimates, a quarter of the world's undiscovered oil and gas reserves lie under the Arctic Ocean, in areas which have historically been seen as undrillable because of thick drifting ice floes. Massive investments are already being made to tap into this economically valuable resource: the Norwegian government is spending billions of dollars building a liquefied natural gas terminal at the far northern port of Hammerfest, whilst a massive gas find in Russian Arctic waters – estimated to contain double Canada's entire reserves – has sparked an unseemly scramble amongst oil majors to partner with Russia's giant Gazprom corporation to exploit it. According to one energy analyst quoted in the *New York Times*, this new Arctic rush is 'the Great Game in a cold climate'.

Internationally, Arctic Ocean nations Canada, Denmark, the United States, Norway and Russia are battling to establish undersea mineral rights to 'their' sections of the seabed. In August 2007 Russian explorers mounted a particularly audacious land grab by piloting a submersible under the ice and planting a rustproof metal flag on the seabed 4,000 metres below the North Pole – thereby claiming the entire area, and its fossil fuel riches, for the motherland. 'We are happy that we placed a Russian flag on the ocean bed, and I don't give a damn what some foreign individuals think about that,' declared expedition leader Artur Chilingarov haughtily as he returned to a hero's welcome – complete with champagne and ranting pro-Kremlin youth groups – in Moscow. Stretching unintended irony to breaking point, the triumphant Chilingarov was then handed a large furry toy polar bear – the emblem of the pro-Putin party United Russia (of which Chilingarov is a parliamentary deputy) – as a military brass band played. The US State Department was studiously unmoved, however: 'A metal flag, a rubber flag or a bedsheet on the ocean floor … doesn't have any legal standing,' a spokesman told Reuters. 'You can't go around the world and just plant flags and say "We're claiming this territory,"' complained an annoyed Canadian foreign minister, accusing Russia of behaving like a fifteenth-century colonial explorer.

Real polar bears can't compete with this kind of remorseless economic imperative. Polar bears – *Ursus maritimus* – have an umbilical relationship with the sea. Adult bears spend most of their lives out on the ice, hunting seals and other prey, and can travel thousands of kilometres each year across the polar seas. They are remarkable animals by any standard: with two layers of fur added to four inches of blubber, polar bears experience almost no heat loss even in the coldest temperatures. This is fortunate given that a bear may have to wait motionless for days beside a seal's breathing hole before getting a five-second chance to swipe one.

Polar bears that are stranded on land in the ice-free summer face lean times, with only berries, seaweed, old carcasses and human refuse to feed on. The earlier the ice breaks up in the spring, the less chance bears have to replenish their reserves before the lean summer months. Already one scientific study has found evidence that Hudson Bay polar bears are in worse condition in years when the sea ice breaks up earlier. Fewer yearling cubs survive, and fewer new cubs are born. The implication is clear: retreating sea ice can only be bad news for polar bears – even as it is good news for oil companies and hard-nosed American entrepreneurs.

Yet less sea ice is exactly what every single forecast now projects, as the previous chapter showed. Models differ about when exactly the permanent summer ice will disappear from the pole, but there is no disagreement about the direction of the change. One study by the NASA scientist Josefino Comiso looked specifically at the amount of ice that might be left in a two-degree world, and concluded that huge areas of ice-free water would open up north of Canada, Alaska and Siberia, perhaps as early as 2025. Year-round ice will still survive, but with less and less area each summer, crowding polar bears into an ever-smaller remnant between the top of Greenland and the North Pole, or leaving them stranded and starving on land. As the 2004 Arctic Climate Impact Assessment concluded ominously: 'It is difficult to envisage the survival of polar bears as a species given a zero summer sea-ice scenario.'

Nor are polar bears the only species to be affected: ringed seals, their major food source, also live their whole lives on or under sea ice. Walruses too need ice fairly close to land where the seas are shallow – they dive from ice platforms and feed on the sea bottom. If the remaining areas of sea ice drift far offshore from the continental shelves, the sea will be too deep for walruses to feed. (Again, these changes are already under way: in Alaska, hunters working in open seas have reported desperate walruses trying to climb on board, mistaking their white boats for vanished ice floes.)

In fact, the whole food web will shift as temperatures rise and sea ice diminishes – from the plankton which are the primary marine producers, through to fish, seabirds and mammals. On land, caribou may starve in large numbers as freezing rain replaces snow, coating the plants they graze on in a thick layer of ice. Several bird species, like the emperor goose, are projected to lose more than half of their habitat. Freshwater fish species like Arctic char, grayling and northern pike will also suffer a decline because of warmer waters. Although warm-loving species will benefit and move north, cold-adapted Arctic animals and plants will find their survival threatened, and extinction will loom.

The landscape itself will change. A recent study simulated the effect of 2°C global warming on vegetation types in the Arctic, and found that tundra almost completely disappears, squeezed closer and closer to the northern coasts of Alaska, Canada and Siberia as the forests march northward. Trees even invade Greenland. Cold moss and lichen tundra is almost driven to extinction, with remnants clinging to life only on the highest mountains and remotest northern islands. The permafrost boundary retreats hundreds of kilometres to the north, destabilising forests, buildings and mountainsides as the ground thaws out.

Again the culprit is the 'Arctic amplifier' of global warming, which means that a temperature rise of two degrees globally would lead to anything from 3.2°C to 6.6°C warming in the Arctic by 2050. The speed of the transition would be *at minimum* half a degree per decade, and at maximum up to 1.5°C per decade. These rapid warming shifts not only outstrip anything witnessed in the region for hundreds of thousands of years, but will also outpace the adaptive capacity of plants, animals and humans alike. All will struggle to survive the new century.

Arctic peoples – being among the very numerous *Homo sapiens* like the rest of us – certainly won't be threatened as a species by the warming climate, but they will find themselves endangered on a cultural level. As one Canadian newspaper remarked, the Inuit may have twenty words for snow, but they're rather short of terms for 'climate change' and 'greenhouse gas'. They do however have a word for crazy weather: *uggianaqtuq*, which translates more or less as 'to behave unexpectedly'.

A lot more than words is at issue, however: ways of life that have depended for thousands of years on the predictable changing of the seasons will be thrown off balance as winter loosens its grip and traditional food supplies disappear. As Chief Gary Harrison told the UN Climate Change Conference in Montreal in December

2005: 'Arctic indigenous peoples are threatened with the extinction or catastrophic decline of entire bird, fish and wildlife populations, including species of caribou, seals, and fish critical to our food security. Climate change threatens to deprive us of our rights, of our rights to sustain ourselves as we have done for thousands of years.'

Unlike animals and plants, however, the Arctic's indigenous human inhabitants can fight back. The Inuit leader Sheila Watt-Cloutier recently lodged a petition with the Inter-American Commission on Human Rights on behalf of all the Inuit people asking for relief 'from violations resulting from global warming caused by acts and omissions of the United States'. Signatories include residents of Shishmaref, the Alaskan village which I visited in *High Tide* to discover how the disappearing sea ice is threatening the survival of the community through higher rates of coastal erosion.

The 130-page document points out that the impacts of climate change 'violate the Inuit's fundamental human rights protected by the American Declaration of the Rights and Duties of Man and other international instruments. These include their rights to the benefits of culture, to property, to the preservation of health, life, physical integrity, security, and a means of subsistence, and to residence, movement, and inviolability of the home.' The Inuit, unlike many around the world who have lost touch with their environment, know what is at stake from global warming.

Indian summer

Unlike the Inuit, however, it is safe to say at present that few people in today's India give much of a damn about global warming. One of the exceptions, the Indian current chair of the Intergovernmental Panel on Climate Change, Rajendra Pachauri, has been a prominent voice raising awareness of the threat of climate change. But even he has to face harsh realities. 'Clearly, at this point of time to do things that are expensive and which will impede our economic growth just to cut down on emissions of greenhouse gases would be something the Indian public would not accept,' Pachauri admitted recently.

One of the few Indians still to quote Gandhi's epithet 'Be the change you want to see in the world', Pachauri is unusual in stopping to think during the headlong national rush to get rich. As the booming middle class buys cars, fridges and air-conditioners in the millions, India's CO_2 emissions are rising by 3 per cent a year, and oil consumption is expected to hit 2.8 million barrels per day by 2010. India overtook Japan as the world's fourth-largest energy consumer in 2001. But such enormous figures don't even make the Indian government break a sweat. 'There is no way that anybody can expect countries like India to cap their emissions for the next 20–25 years,' S. K. Joshi, a senior official in the environment ministry, told Reuters recently.

Joshi must have missed the September 2005 meeting, organised by his own ministry together with Britain's Department for Environment, Food and Rural Affairs, which outlined some of the projected impacts of global warming on India. Studies commissioned by both governments project decreases in agricultural output of both wheat and rice, particularly in the majority of farmland which is not irrigated, with two degrees of warming. Forest types will shift too, with large-scale die-back expected, particularly in traditional forest savannahs. This would have dire implications for the 200,000 villages located in or near forests and heavily dependent on them for people's livelihoods.

The research confirms an earlier study, published in 2001, which projected major agricultural impacts on India for two degrees of global warming. Worst hit would be wheat production in the northern states of Haryana, Punjab and western Uttar Pradesh. Some states like West Bengal would gain – though to a lesser extent, and overall the country would lose 8 per cent of net revenue as a result. For a country with a growing population, the implications for food security would be serious.

These changes will be aggravated by wider meteorological shifts also associated with the rising temperatures, including a strengthening of the monsoon and increasingly severe flooding. India may also suffer refugee influxes as its neighbours face their own climatic problems. Highly populated neighbouring Bangladesh will suffer disproportionately from the stronger monsoon. The country already receives nearly two and a half metres of rainfall each year, flooding 30–70 per cent of its land area even under today's 'normal' climate. With heavier rainfall in stronger storms, millions more people will be displaced by monsoon floods – perhaps permanently if their homes are washed away repeatedly, forcing them to move elsewhere.

In India the overall level of rainfall received during the monsoon season – despite high variability between years – has remained about the same for the last 50 years, according to scientific studies. But global warming is having an effect – by increasing the frequency of intense downpours and reducing the number of gentler rain events during the monsoon season. With more of the year's precipitation coming in torrential thunderstorms, the incidence of severe flooding is already on the rise, and will climb still further in coming decades. This kind of rain sees the heavens open with very little warning, and flash floods spark landslides and wash away homes and villages as rivers burst their banks within minutes of the cloudburst beginning. With very little warning of impending disaster, people have less time to get themselves and their families out of harm's way – and the flooding death toll may rise too as global warming accelerates.

In Nepal too much water will also be a problem, though this water comes from a very different source. As the mighty glaciers of the high Himalayas melt, the run-off they produce tends to pond up behind walls of moraine rubble left by the retreating ice, forming legions of new glacial lakes. These lakes hold huge amounts of water behind their unstable natural dam walls, and breaches can trigger catastrophic mudflows, which hurtle down river valleys, sometimes for distances of up to 200 kilometres, wiping out everything in their path. In 1985 a 10-metre-high wall of water from a glacial lake burst swept down the Bhote Koshi and Dudh Koshi rivers, destroying a hydroelectric plant, 14 bridges, 30 houses and even undermining the Everest airstrip at Lukla. A survey in 2000 found 20 dangerous lakes with the potential to flood out at any time, and in the two-degree world their number will increase substantially.

The meltdown in the mountains will have a more insidious but far worse effect in the longer term, however. As the glaciers disappear altogether from all but the highest peaks, their run-off will cease to power the massive rivers that deliver vital fresh water to hundreds of millions of inhabitants of the Indian subcontinent. Water shortages and famine will be the result, destabilising the entire region, as the next chapter shows. And this time the epicentre of the disaster won't be India, Nepal or Bangladesh – but nuclear-armed Pakistan.

Peru's melting point

Because of their sheer scale and height, the mighty Himalayan glaciers are not about to disappear entirely just yet. The same cannot be said for the more vulnerable Andean ice fields, however. Here global warming has already reduced the area covered by glaciers by a quarter in the last three decades. Lonnie Thompson, the pioneering climatologist who drilled ice cores on the summit of Kilimanjaro, also spent many years drilling cores on the unique Quelccaya ice cap in Peru's eastern mountains. When he first drilled through the ice in 1976, Thompson found clear annual layers going back in time for 1,500 years. But when he returned in the early 1990s, the topmost and most recent layers had already been destroyed by percolating meltwater. He concluded sadly that a mountain fastness which had remained permanently frozen for over a millennium was now melting down.

Peru's capital Lima doesn't look like a city which depends on the highlands for its fresh water. Located on the country's barren coastal strip, its nearest mountains are utterly desolate wastelands of rock and sand, supporting not a single blade of grass, never mind a whole city. So how does Lima survive, in one of the driest desert areas in the world? Not from rainfall: the paltry 23 mm annual average precipitation comes exclusively in the form of a chilly drizzle, which immediately evaporates as soon as the sun comes out. Instead, the answer emerges if you follow the main highway inland, along the deep valley of the river Rimac. The sun-baked desert mountains soon rise into grass- and then snow-covered peaks, some of them towering over five and a half thousand metres in height – high enough to give me a near-death experience of acute altitude sickness when I approached them incautiously, as I related in *High Tide*. These glaciated mountains are Lima's water towers – natural reservoirs in the sky. They keep the Rimac River flowing during the Andean dry season when even the high mountains see little rain or snowfall.

Strangely, there have been no scientific studies conducted to my knowledge on the likely impact of two or more degrees of global warming on Lima's life-supporting glaciers. However, other river catchments in the Peruvian Andes have been studied closely. The Rio Santa, which drains the Cordillera Blanca chain of summits (including the 6,768-metre Huascarán, Peru's highest peak, and the scene of another of Lonnie Thompson's ice-drilling escapades), depends almost entirely for its dry-season run-off on glacial melt, according to a joint Austrian–Peruvian hydrological study. Downstream from the high mountains, the Rio Santa's water is channelled through hydroelectric turbines – in the spectacular Canyon del Pato – to produce 5 per cent of the entire country's electricity, and its waters also support vast green fields of maize, melons and sugar cane on the otherwise arid coastal plain. The inhabitants of the coastal cities of Chimbote and Trujillo – the latter is home to more than a million people – also depend on the Santa for their drinking water.

Yet by 2050 the Cordillera Blanca's glaciers will have shrunk by 40–60 per cent, according to a second study by two of the same authors. Their model predicts a fall in glacial run-off of nearly half by the same date, which will reduce flows drastically during the dry season. By this stage the Peruvian authorities will be faced with some difficult choices. Should they release water into the river from the few managed lakes in order to keep hydropower production stable? Or should water be retained in order to keep supplies to the cities from running out? Agricultural production will be dwindling by this time, with huge increases in unemployment as the irrigated coastal fields return to the desert from whence they came.

Peru's ancient history may conceal a lesson or two along these lines. Further north along the coast from Trujillo lies the Jequetepeque valley, site of the major pre-Columbian civilisations Moche (AD 200–800) and Chimu (AD 1100–1470), both of which cultivated extensive areas on the now barren valley floor. They also built networks of irrigation canals and aqueducts to bring scarce water from streams far away in the mountains, and constructed windbreaks in the desert to stop migrating dunes from inundating their fields and houses. Both societies managed to cope with floods resulting from the periodic visits of El Niño, and even learned to grab a harvest or two from the wet patches the floodwaters left behind.

One thing they couldn't cope with was drought. The upper reaches of the Jequetepeque River have no glaciers: its flow is therefore completely dependent on a regular supply of water from highland rainfall. Some of its dry-season flow is buffered by small lakes and wetlands, which like ice tend to release water slowly. But if the rains fail for too long, then the river dries up, leaving people with no water for drinking or irrigating their crops. There are tell-tale signs, uncovered recently by archaeologists, that the periodic crises which engulfed the Chimu and Moche peoples were directly linked to droughts. When the rains failed and the rivers dried up, all hell broke loose. The result was an unpleasant mess of warfare, migration and eventual societal collapse.

Today's urban societies are of course very different from those constructed by the Moche and Chimu. At its height, the Moche's entire civilisation included only half a million or so people, scattered across several coastal valleys. Today Lima has upwards of 8 million (a third of Peru's entire population) over a vast area, many of them eking out a living in the impoverished shanty towns which sprawl up the hillsides around the city. The Moche may have had canals and aqueducts, but they could never have dreamt of the kind of water infrastructure that is now considered normal in a modern-day city. When I was in Peru in 2002, I visited the Lima water utility SEDAPAL's space-age buildings on the outskirts of town. Despite perennial funding problems, SEDAPAL's operations were humming along as normal, with huge undercover reservoirs, looking like Olympic swimming pools, channelling crystal-clear water to different parts of the city.

But SEDAPAL has a problem. The inland Cordillera Central, on which the river Rimac depends for glacial run-off, is lower than the more extensively glaciated Cordillera Blanca, with even the highest peaks barely crossing 5,500 metres in altitude. The glaciers are small, without extensive snowfields to feed them. SEDAPAL has built reservoirs to catch wet-season rainfall, but available sites – and funding – are limited.

It's difficult to envisage the survival of any glaciers in the Cordillera Central in the two-degree world. Even the crudest back-of-the-envelope calculation indicates that trouble is brewing. A rule of thumb is that each degree rise in temperature raises freezing levels by 150 metres, so if the global average is extrapolated to the Andean mountains, freezing levels – and therefore glaciers – will have retreated 300 metres higher up the slopes. Glaciers are currently only found above 5,000 metres. With ice only surviving then above 5,300 metres, only tiny patches will be left on the highest peaks by the time world temperatures reach two degrees. Actually, the situation is worse than this crude calculation suggests, because temperature rises are higher in mountain regions than the global average thanks to a quirk of atmospheric physics which means that temperature rises are faster the higher you go in the atmosphere. Indeed, the temperature in the tropical Andes is already rising at twice the global rate, adding to the rapid meltdown which is already under way. There is only one conclusion: Lima's natural water towers are doomed to dry up.

Moreover, Lima is not the only major city in the region to be heavily dependent on glacial meltwater: Peru's neighbours Ecuador and Bolivia also survive thanks to run-off from their mountain glaciers. Ecuador's capital Quito currently gets part of its drinking water from a glacier on a nearby volcano called Antisana – where, as elsewhere in the Andes, the ice is already retreating fast. All these nations also rely on mountain water for hydroelectricity production – and as river flows falter during future dry seasons, either the population will suffer long-lasting failures in power supplies, or alternative sources must be found, perhaps using fossil fuels.

So how will the inhabitants of a city like Lima cope? Experience suggests that the poorest will be hit first. Rich people might be able to afford expensive bottled water trucked in from afar, piped through the mountains or perhaps squeezed out of desalination plants whose output would be too costly for the poor. Agriculture will also suffer, putting hundreds of thousands out of work up and down the Peruvian coast. In a chronic water shortage situation, Lima's streets might begin to empty, in a strange kind of reverse migration – instead of people moving from the countryside to the city, people might make the trek back to their mountain villages, where water supplies are more plentiful and crops can still be grown. The city's influence would dwindle, and the half of Peru's population that currently lives in the desert might be forced to move up into the mountains – assuming space and cultivable land can be found for them. Major refugee upheavals seem likely, with widespread conflicts between water users as river flows slow to a trickle. The coastal human geography of Peru will look very different once the high Andean glaciers are no more.

Sun and snow in California

With no major glaciers outside Alaska, the United States might imagine itself immune to the water crisis bearing down on Peru. It could not be more wrong. Towns and cities all the way up America's west coast are heavily dependent on frozen mountain water – not from glacial ice this time, but from winter snows. In the great river basins of California, Washington and Oregon, far more water is stored in snowpack during the spring and early summer than in man-made reservoirs behind dams. Snowpack acts as a natural reservoir, holding winter precipitation and releasing it slowly during the drier months of the year as the snow gradually melts. But as the global and regional temperature climbs higher in decades to come, winter snow will increasingly be replaced by rain in the Sierra Nevada, Cascades and Rocky Mountains. This isn't just bad news for the ski industry, it's bad news for anyone who wants to turn on their tap in the summer and find fresh water flowing out of it.

Even in today's climate, water resources in the arid west are stretched to breaking point. With all its dams and irrigation canals, the Colorado River has been sucked dry by the time it reaches the Gulf of California – most of the time, no water reaches the sea at all. (A once productive wetland ecosystem of marshes, fish and bird life has been all but destroyed as a result.) The Colorado's water isn't just used to irrigate golf courses in Las Vegas; it also supplies drinking water and hydroelectric power to much of southern California and Arizona. Battles have broken out between different states: in August 2005 protesters in Salt Lake City took to the streets to oppose a plan by the Southern Nevada Water Authority to pump groundwater through 500 miles of pipes south to Las Vegas.

The San Joaquin River, its waters diverted to feed the fertile fields of California's productive Central Valley, also mostly fails to reach the sea down its old natural river bed. As an Associated Press reporter put it: 'Where spawning Chinook salmon once ran thick, lizards and tumbleweed inhabit a riverbed that often goes for years without water.' Instead of sustaining salmon, the river now sustains oranges: 80 per cent of America's eating oranges are grown in the state. With major cities like Los Angeles and Sacramento all dependent on a network of canals and pipelines for their water, the hydrology of California is more like a giant plumbing system than a set of natural river valleys.

But however resourceful the state's water engineers might be, they are going to struggle to find a way out of the collision course between a burgeoning population and a declining water supply as the world warms. One major recent study, published in the journal *Proceedings of the National Academy of Sciences* in 2004, projected snowpack declines of between a third and three-quarters in the two-degree world. In addition, heatwaves in Los Angeles will quadruple in frequency, whilst crippling droughts will occur 50 per cent more often, increasing the demand for scarce water. With Sierra Nevada snowpack declining and earlier melt producing earlier spring run-off, surface water supplies to 85 per cent of Californians – orange growers and city dwellers alike – will be reduced.

Nor would the changes be limited to California: a second study projects snowpack declines in Oregon and Washington too, with the northern Rockies and Cascades seeing declines of 20 to 70 per cent. With more rain and earlier snowmelt, the likelihood of winter flooding will also increase, even whilst reduced run-off causes water shortages in the summer. Most affected by increased floods will be mountainous regions of the California Coastal Range and the Sierra Nevada. Further north, in the Columbia River system, earlier run-off peaks mean that water managers will have to choose between keeping water in reservoirs for hydropower production during the summer, or releasing it earlier in the year to prevent salmon being driven to extinction. In this unpleasant choice between keeping the lights on or the fish happy, it's not difficult to imagine which way most people would vote.

In addition, summer droughts mean forests becoming tinderdry and increasingly susceptible to wildfires. The northern Rockies, Great Basin and Southwest could see the length of their fire season – which already exacts a deadly toll in bad years – extended by two to three weeks. The study overview concludes ominously: 'Current demands on water resources in many parts of the West will not be met under plausible future climate conditions, much less the demands of a larger population and larger economy.'

The previous chapter saw how the western interior of the United States, from Nebraska down to Texas, could be facing a major drought disaster many times worse than the 1930s Dust Bowl with only small additional increases in global temperature. Back in the Dust Bowl years, Pacific states like California were a refuge for those displaced by the drought. But in the two-degree world, these west coast havens will be facing critical water short-ages of their own. The changes in snowpack and run-off will not just mean that golf courses and ski resorts bite the dust. They call into question the whole region's capacity to support big cities and agricultural areas. California will no longer be the golden state once global warming begins to bite.

Feeding the eight billion

However serious the water crisis, it is highly unlikely that anyone will starve in the western US with two degrees of global warming. Some areas of the continent may even benefit from better growing conditions. Moreover, as the richest country in the world, America's purchasing power in world food markets will ensure that its citizens survive even the toughest droughts for the foresee-able future.

That is only true of course as long as there is food on the world markets to buy – and the two-degree world will see escalating challenges as crop-producing areas struggle to adjust to a warming climate. Meantime, those countries which suffer most, and don't have massive purchasing power, will slip irrevocably into structural famine and crisis. By a cruel irony, these include many of the countries that have done least to cause climate change.

There is some good news, however. The northern central part of the US, including such states as Indiana, Illinois, Ohio, Michigan and Wisconsin, will become an increasingly important winter wheat-producing area. Currently the crop is only marginal due to low cold-season temperatures, but with warmer winters yields are expected to double. North Dakota and Minnesota will see jumps in maize production, whilst the whole northern section of the country can expect an increase in potato production as long as rainfall remains reliable. Citrus growers in states like Florida – which in the current climate can see big losses due to occasional freezing temperatures – will gain too through a reduction in frosts.

Across the Atlantic, the UK won't yet be growing lemons, but will see increasingly successful crops of sweetcorn, soft fruits like strawberries, and popular vegetables like onions and courgettes. As winters get milder and wetter, winter wheat will thrive along with warm-loving legumes like navy beans (the traditional baked bean), which could become a major crop in southern England. Many of these crops may also benefit from the fertilising effect of an atmosphere richer in carbon dioxide, and across Europe as a whole, warm-loving crops like sunflowers and soybeans will be able to grow much further north. Wheat and maize may even expand into new areas of western Russia and southern Scandinavia.

Maize is one of the world's great staple crops, essential for household income and food security in many developing countries. This is where the problems begin. In Central and South America, the region where maize was first cultivated for food by the ancient Maya, losses are projected in every country except Chile and Ecuador. These losses may be offset by technological improvements in the future, but subsistence family farmers will be less able to adapt than big mechanised growers.

The majority of Africa is also expected to experience big

declines in yields. In 29 African countries the risk of crop failure and hunger is projected to increase, with the worst-hit including nations already vulnerable to famine such as Burkina Faso, Swaziland, Gabon and Zimbabwe. Only highland African countries like Lesotho and Ethiopia can expect increases in yields in areas that are currently too cold to support much maize production. In Mali, up to three-quarters of the population could be at risk of hunger with climate change, up from a third today, whilst in Botswana up to a third of the maize and sorghum crop could be wiped out due to declining rainfall. In the Congo, which lies in the equatorial belt and can expect increases in rainfall in a warming world, food supplies could still decline because farmers need a good dry season to burn off land for next year's crops.

The US will also be affected. In south-eastern America, lower summer rainfall and soaring temperatures could slash soybean production by half, with similar yield decreases projected for sorghum. And although higher temperatures will open up new agricultural opportunities in Canada, traditional food sources – from sugar maples to salmon – will lose out in a two-degree world. Fish populations will also suffer on both sides of the Atlantic: whilst salmon populations will diminish in Canada, North Sea cod will be virtually wiped out because of warming waters unless a total ban on fishing is introduced.

In the unemotional lottery of global warming, whether you eat will depend on where you live. Inhabit a rich country with reliable rainfall, and you won't go hungry – even if traditional meals like fish and chips have to change with the times. But live in the drier subtropics and life will become increasingly precarious.

Of course, famines already occur in today's world, even though there is enough food for everyone in a global sense. As aid agencies frequently complain, the issue is poverty, not just drought. But if one thing is certain, it is that famines are more likely in a world with less food to go round overall. With increasing competition over diminishing harvests, prices will soar on world markets during lean years. Food price stability in the two-degree world will depend on northern land areas being opened up to new crops rapidly enough to replace yield losses in hotter, drier areas to the south.

With assiduous planning and adaptation, the world need not tip into serious food deficit. However, if the temperature rises past two degrees, preventing mass starvation will be increasingly difficult, as future chapters will show. First millions, then billions of people will face an increasingly tough battle to survive as rising temperatures make growing sufficient food an ever more difficult task.

Silent summer

Most readers will by now have concluded that two degrees of global warming is likely to be survivable – barring any unpleasant surprises – for the majority of humanity. I agree. The same cannot be said, unfortunately, for a large swathe of natural biodiversity: the plants and animals that share this planet with us. With ecosystems already fragmented and marginalised due to incessant human population growth and economic activity, climate change looks set to reap a grim toll on what remains of nature.

Threats to some emblematic species, like the pikas, were mentioned in the previous chapter, as were especially vulnerable ecosystems like coral reefs and the Queensland Wet Tropics rainforest. All will be suffering further serious reductions in the two-degree world. South Africa's majestic protea flowers will be losing much of their range, and 10 per cent could be extinct by 2050. In the Queensland rainforest, around a third of all the species studied will be well on their way to obliteration as temperatures hit two degrees. In each of these cases an outcry can be expected as well-loved animals and plants die off. But who will stop to shed a tear for the Monteverde harlequin frog of Costa Rica, the Canadian collared lemming, the Hawaiian honeycreeper or any of the other thousands of lesser-known species whose survival is threatened as global warming leaves them stranded in a climate zone no longer appropriate to their needs?

Given the pressure that nature is already under from human activities, climate change could not have come at a worse time. Already – independently from any change in the world's climate – we are living through what biologists have termed the sixth mass extinction of life on Earth (the fifth was the extinction of the dinosaurs and half of all other life at the Cretaceous–Tertiary boundary). Due to combined human pressures from habitat loss, hunting, pollution, resource use and the introduction of invasive species into new areas, natural species are already becoming extinct at a rate 100–1,000 times greater than the normal background rate of loss over evolutionary time. The most comprehensive-ever survey of the planet's health, the UN Millennium Ecosystem Assessment, brought together 1,360 experts from 95 countries, who concluded that a full two-thirds of the ecosystems humans depend on are currently being degraded or used unsustainably.

It is also clear from many different broad-based studies that nature is already being affected by global warming. One looked at 100 species across the globe and found an average move of 6 kilometres towards the poles (and 6 metres up mountains) per decade. Some butterflies have already moved their ranges by 200 kilometres. Spring is coming earlier too – birds are laying, tree buds bursting and frogs spawning earlier in the year. A second global analysis identified clear global-warming-related changes in range and behaviour with hundreds of species from molluscs to mammals and grasses to trees.

Many of us will have begun to notice these changes in our own local environments. Apple trees in my local community orchard, for example, were inadvisably bursting into blossom in late October in 2006 after unseasonably warm temperatures tricked them into thinking spring had already arrived. I have also noticed a strange absence of bird life – for several months the garden has been nearly silent. The English weather has changed very dramatically in just a few years: I have childhood memories of hard frosts on Halloween, for example, but this year temperatures had still not dipped below freezing even when November came around. The trees in the woodland which divides our property from the nearby Oxford Canal stayed resolutely dark green until two months after autumn colours should have begun to show. Almost every reader will now be able to add anecdotal observations of the same sort.

Of crucial importance for natural species is the speed of temperature rise. A 2°C increase occurring over 1,000 years would only mean a 0.02°C increase per decade, a slow rate of change that most species – all other things being equal – would have a high chance of adapting to. But if the same increase took place over 50 years (a 0.4°C increase per decade, and a far more likely scenario) the impact would be catastrophic. Whilst you might think that mobile species like butterflies could move easily in response to temperature changes, studies by Professor Jeremy Thomas on the English blue butterfly found that it can only disperse to new areas at the glacial rate of 2 kilometres per decade, whilst other insects like ground beetles moved at only a tenth of that rate. Faced with the need to move at 30 kilometres per decade – or 3 kilometres per year – to keep up with shifting temperature zones, these sedentary species clearly have a problem. So do rooted plants, of course, which can't physically move at all, and therefore have to rely on seed dispersal for gradual population movement. The seed dispersal limit for many forest trees is less than a kilometre a year. The forest cannot move any faster, and will soon be outrun by a rapidly changing climate.

Temperature rise is not the only variable, of course – drought more than heat is the biggest threat to beech woodlands near my home in Oxford. As ecosystems unravel, species which are finely attuned to each other will be thrown out of sync. In Holland, populations of pied flycatchers have declined by 90 per cent over the last two decades, because their chicks' hatching is currently mistimed with the advancing spring. By the time hungry nestlings need most food, the populations of caterpillars they depend on have already peaked – and the baby birds slowly starve to death.

Species have evolved to fill particular ecological niches, which may disappear as other species die out or migrate. Animals and plants also tend to be highly adapted to their geographical habitat. Chalk grasslands, for example, will not have much success moving north if the soils in cooler climes are all underlain by clay or granite. Habitat fragmentation is another problem: cities, agricultural monocrop 'deserts' and major roads all present insurmountable barriers to species migration. In southern England, the timid dormouse will not cross open fields, let alone scurry through the busy streets of Birmingham on its supposed journey north. As a result, climate change calls into question the very basis of site-based nature conservation: there is no point in declaring somewhere a nature reserve if all the species within it have to flee north within a few decades in order to avoid going extinct.

All these concerns focus on the climatic 'envelope' that species inhabit, and this also provides the approach for what may well turn out to be one of the most important scientific papers ever written. In a study published in *Nature* in 2004, the ecologist Chris Thomas and more than a dozen other experts revealed that according to their models over a third of all species would be 'committed to extinction' by the time global temperatures reached two degrees in 2050. 'Well over a million species could be threatened with extinction as a result of climate change,' Thomas told the press.

Some of these would include the iridescent blue western jewel butterfly, native to south-western Australia; Boyd's forest dragon, a spectacular crested lizard from the threatened Queensland Wet Tropics; half the 163 tree species currently found on the Brazilian Cerrado savannah; between 11 and 17 per cent of all European plants and a quarter of the continent's birds, with the red kite, dunnock, crested tit, Scottish crossbill and spotless starling near the top of the list; the smokey pocket gopher and Jico deer mouse from Mexico's flat Chihuahuan Desert; and 60 per cent of the species currently resident in South Africa's famed Kruger National Park – to name but a few.

It is worth pausing for a minute to take in the full significance of this projected global cull. In a chilling echo of more innocent times, Thomas and his colleagues' paper reminds me in reverse of Charles Darwin's *The Origin of Species*. Whilst Darwin's work laid out the theory for species' evolution down the ages, the 2004 *Nature* paper maps out their projected disappearance. Were Darwin to have written the study today, he might have called it *The End of Species*.

Consider the thought that living species, which have evolved on this planet over millions of years, could be destroyed for ever in the space of one human generation; that life, in all its fascinating exuberance, can be erased so quickly, and with such leaden finality. As the biologist Edward O. Wilson has suggested, the next century could be an 'Age of Loneliness', when humanity finds itself nearly alone on a devastated planet. In tribute to Rachel Carson, I call this our Silent Summer – a never-ending heatwave, devoid of birdsong, insect hum, and all the weird and wonderful living noises that subconsciously keep us company.

Should we just accept this fate? asks Wilson. Should we knowingly erase Earth's living history? 'Then also burn the libraries and art galleries,' he demands, 'make cordwood of the musical instruments, pulp the musical scores, erase Shakespeare, Beethoven, and Goethe, and the Beatles too, because all these – or at least fairly good substitutes – can be re-created.' Unlike, of course, Boyd's forest dragon or the golden toad – the former under great danger, the latter already gone for ever, thanks to climate change.

Nor is the loss of biodiversity just an aesthetic concern. Whilst I and many people feel that natural life and biodiversity have an intrinsic value, separate from their use to humans, all of human society is at root dependent on natural ecosystems. This might come as news to the average city dweller tucking into a ready meal in front of the TV, but it doesn't make it any less true. From fish to fuel wood, nature's bounty feeds us, houses us, warms us and clothes us. Soils wouldn't support agriculture were it not for the organic matter broken down by bacteria. Crops wouldn't set seed unless pollinated by bees. The air wouldn't be breathable were it not for photosynthesis by trees and plankton. Water wouldn't be drinkable were it not for the cleansing action of forests and wetlands. Many of the medicines that extend our lifespans were first developed from natural substances produced by plants and animals, and many more undoubtedly remain to be discovered. Life even regulates the nutrient cycles of the planet: had ocean-dwelling organisms not sequestered excess carbon into limestone and chalk over millions of years, our habitable planet would long ago have turned into Venus, which suffers blistering surface temperatures of 500°C – hot enough to melt lead – thanks to an inhospitable atmosphere composed 96 per cent of carbon dioxide.

Some of these ecosystem services can be replaced by technology, as many economists might suggest. Think, for example, of hydroponics: the replacing of natural soil with synthetic rooting material and a cocktail of chemicals. But ecology is such a complicated web that we cannot even understand many of the living interactions that go on within ecosystems, let alone imagine that we can somehow redesign and replace them. Scientists once tried to build a sealed living world – nicknamed Biosphere 2 – from scratch in a big greenhouse in the Arizona desert. They failed. As carbon dioxide levels rose within the sealed greenhouses, Biosphere 2's human inhabitants must have reflected on the lessons they were learning as they gasped for air. Functioning ecosystems cannot be created artificially. Life keeps us alive, and we lay waste to it at our peril.

3°

3

THREE DEGREES

What every Botswanan wants

Botswanans have a major national obsession. It isn't dancing: the rather drab capital Gaborone is not known for its wild parties or nightlife. Nor is it sport – Botswanans excel at neither football nor athletics, and patriotic citizens often lament that the country's national anthem has still never been heard at the medal-giving ceremony of any major sporting competition. This national obsession has a long heritage, however – indeed it was once a key element of pre-Christian religion. It is still evident in daily conversation right across the country, and in the habit many people have of staring wistfully at the horizon on the hottest summer days in January and February. It's an obsession which turned the flag blue and gave the national currency its name. It's rain.

Unlike the English (though that may change), Botswanans love rain. The Setswana greeting *pula* literally wishes for rain, in the same way that the Hebrew *shalom* wishes for peace. In their hot, dusty and largely flat land, Botswanans live for the day when rumbling clouds gather, the heavens open, and fat raindrops splatter onto the sun-baked earth. When Botswana (formerly Bechuanaland) became independent from Britain in 1966, its first

101

president, the still-revered Sir Seretse Khama, punched the air and shouted: 'Let there be rain!' Rarely can there have been a clearer articulation of national will.

Though a relative African success story, Botswana is a country on the very margins of existence. Less than 1 per cent of the land surface is officially considered cultivable, and the Kalahari Desert dominates the south and west of the country. Rather than rising in Botswana, the Okavango River ends there, dying gradually under the hot sun in a flat delta of salt pans and wildlife-rich wetlands. Though diamonds are the mainstay of the economy (along with tourism), cattle are central to agricultural and cultural life. Owning ten thousand head of cattle would impress Botswanans far more than the possession of a posh condominium on New York's Fifth Avenue.

But sadly for Botswana, the long-range forecast shows very little rain. By the time global warming reaches three degrees, drought will have already become perennial in both this country and much of the rest of southern Africa. Even whilst tropical areas and the higher mid-latitudes drown in floods, the subtropics will be simply baking to death. The culprit is not, as one might expect, the parched land. It is the sea. The Indian Ocean to the east is already warming rapidly, and is implicated in the devastating drought which has struck southern Africa in recent years. The problem is this: rain clouds form over the warm ocean, and instead of drifting inland to relieve the summer misery of Botswana and its neighbours, they pour their torrential down-pours uselessly back onto the ocean surface. Instead of getting much-needed rain, southern Africa gets the fallout – sinking air which rose in thunderstorms over the Indian Ocean and has now had all the moisture squeezed out of it.

This mechanism is confirmed by computer models of current and likely future climate. A team led by Martin Hoerling of the US

National Oceanic and Atmospheric Administration found that Africa was literally split in half by global warming. The northern half is likely to see a recovery in rainfall, whilst the southern half gets progressively drier. Hoerling's team did not use just one computer model to get this result: they saw the same mechanism repeatedly take place in sixty different simulations using five different models, making their prediction very robust in scientific terms. Even as early as 2010, southern Africa stays consistently dry in the projections, eventually losing 10–20 per cent of its rainfall. A grim forecast indeed.

Cynics might say that sub-Saharan Africans are well accustomed to drought. But the evidence suggests that the extent of drying in the three degrees world is going to be far off any scale that would permit human adaptation. And for people already eking out a living on the margins of subsistence, the result can be summed up in one word: famine.

Exactly how and where this disaster will unfold is not something which is predictable with any confidence. But a separate study suggests that Botswana in particular will be at the epicentre of it. Moreover, neighbouring countries will be in no position to absorb large numbers of Botswanan refugees, because they too will be severely affected. The reason is simple: with sufficient global warming the Kalahari Desert – which currently supports savannah and scrub over large areas – becomes a true hyper-arid desert once again, complete with raging sandstorms and rapidly diminishing vegetation.

Like the American High Plains, much of the Kalahari region currently consists of 'stabilised' dunefields – sand seas which long since stopped blowing, and which support large areas of pastoralist cattle herding and subsistence agriculture. Millions of people provide for themselves and their families across the area by growing staples like sorghum, millet, pumpkins and maize in small, hand-tended fields. To the untrained eye the dormant dunes are not immediately obvious; the landscape simply looks like gently rolling hills, like a browner version of the English South Downs. But each small hill and valley is a linear dune, a single landform that can be tens of kilometres long, part of a sea of dunes which – like waves on the ocean – marches across thousands of square kilometres of landscape. These stabilised sand dunes cover vast expanses: the Northern Kalahari Dunefield extends into modern-day Zambia, Angola and Namibia, whilst the Eastern Kalahari Dunefield encompasses large areas of western Zimbabwe. The Southern Dunefield, meanwhile, extends all the way down to the northernmost regions of South Africa. And stuck right in the middle, of course, is Botswana.

There are two reasons why the dunes lie idle today. First, rainfall – limited though it might be – is still sufficient to support vegetation. Second, wind speeds are low enough that sand can't blow any great distance, and the giant dunes stay put. However, the computer model study led by Oxford University's David Thomas forecasts that the great sand seas of the Kalahari will be well on their way to total remobilisation once global warming reaches three degrees. Again, several different computer models were used, and tested initially to check their ability to successfully 'hindcast' the observed 1961–90 regional climate. All passed the test.

Although the models do not make a perfect match with Hoerling's study, they do suggest large-scale drying – not because of the Indian Ocean this time, but because hotter temperatures across the region increase the amount of water evaporating from the land surface and from vegetation. On its own, this change might not be enough to remobilise the Kalahari dunefields. But the models also project large increases in wind speeds – up to a doubling by 2040 – which will multiply the erosive force to which the old dunes are subjected. Worst-hit will be the Southern

Dunefield, but after 2040 the northern and eastern areas – encompassing Botswana, Namibia, Angola, Zimbabwe and Zambia – will likely all see increasing dune activity, well before global temperatures hit three degrees.

Once this global warming threshold is reached, the models project that little else will remain on the Kalahari but violently blowing sand. With soaring temperatures and howling winds, colossal storms will shift immense quantities of sand and dust across the region, building new dunes and blotting out villages, towns and even entire capital cities like Gaborone. Even at the coarse resolution of a computer model grid, Botswana's fate is clear – the entire country is covered by 'active' dunes after about 2070.

As Thomas and his team note with concern: 'These would represent considerable, even catastrophic, limitations on present agricultural uses of these environments.' In other words, much of the area will no longer be able to support human habitation. Botswana as we know it will drown – not under water, but sand.

Perils of the Pliocene

Sceptical readers might question whether it is sensible to put too much faith in modelling studies, when by definition they can only be as good as their human designers. Even when run on supercomputers like Japan's Earth Simulator – a machine the size of four tennis courts, which can process 35 trillion calculations a second – models still can't accurately replicate every small-scale reality of how the atmosphere works, because of its sheer complexity. The Hadley Centre's HadCM3 model, for example, divides the atmosphere up into a global grid of 9,673 cells, producing boxes of about 300 square kilometres at the mid-latitudes. This can barely resolve the British Isles, let alone the Scottish

Highlands, and still less the Cairngorms. Its representation of changes in mountain-induced rainfall in Scotland will therefore be somewhat less than accurate.

One solution is to embed a regional climate model with much higher resolution inside the global model. Many of the papers cited so far in this book operate on this principle. In order to accurately resolve hurricanes, for instance, which have powerful eyes commonly only a few kilometres across, modellers need to 'nest' a regional model (with a high-resolution grid spacing of only 9 kilometres) within a global model.

But as far as broad-brush stuff like calculating global average temperatures is concerned, the current models do a pretty good job. Climate change over the past century can be simulated with almost unerring accuracy by the most powerful recent models, suggesting that the equations governing the atmosphere's response to the heat-trapping effects of greenhouse gases are now pretty close to reality. And model evaluation is based on much more than guesswork. 'Hindcasting' the twentieth century is one way of checking that a model works, but models can also be calibrated using other time periods – the depths of the last ice age, for example, or warmer periods in the more distant past. After all, to be believable as a predictor of the future a model needs to be able to accurately simulate the past.

Nevertheless, many sceptics base their objections on the suspicion that models are somehow fiddled in advance to come up with the 'right answers' by scientists eager for the next global warming grant – 'you get out what you put in', as the old adage goes. But climate models do have an important grounding: they are based not on subjective judgements by their constructors but on the fundamental laws of physics. These observable physical laws, governing everything from convection within clouds to the reflectivity of sea ice, cannot be changed by anyone, whatever their politics. After all, models don't do anything magical. All they do is solve physical equations. All the processes of HadCM3, for instance, could theoretically be worked out by hand – except that it would then take centuries of human labour to complete one 'model run'. What computers do is speed up the process, just as pocket calculators speed up mathematics lessons in school.

No one, however, suggests that models are perfect. They all tend to come out with slightly different answers to the same question, a reflection of their varying design. The reason here is that some of the physical laws which underpin them are not known precisely. How clouds interact with the wider atmosphere is a big uncertainty, for example, so some cloud model parameters are best guesses. Nor is it known exactly how far sulphate 'aerosols' – tiny particles of pollution blamed for 'global dimming' – cool things down. But models are a useful tool, and give a valuable insight into likely future conditions on this planet – something humanity has never had access to before. Unlike the oracles consulted by the ancients, models offer a way of divining the future based not on the miraculous visions of some unseen prophetess but on observable physical data.

Apart from modelling, there is another option for looking forwards into the future – and that is looking backwards into the past. Many of the case studies explored so far in this book have been based on palaeoclimate research, where warmer periods in the Earth's history can be useful analogues for what might come during this century. Previous chapters have looked back to the early Holocene, less than 10,000 years ago, and the last interglacial, around 130,000 years ago. For an analogue for the three degrees world, we have to go back much further, before the Earth entered its regular cycles of ice ages and interglacials. We have to go back a full 3 million years, to a period of time called the Pliocene.

The Pliocene epoch is of particular interest because it was in many ways quite similar to the world we live in now. Global geography was pretty much as it is today. Great mountain chains like the Andes and the Himalayas already existed at near their present height, and the Isthmus of Panama had recently closed, cutting off the mid-Atlantic from the Pacific and setting up ocean circulation patterns which are still in existence now. The British Isles were even separated from the Continent by the Channel, just as they are today.

There were no humans living in Britain, however. Our primate ancestors were still exclusively in Africa – indeed, the famous 'Lucy' hominid fossil found in Ethiopia is of Pliocene age. She and other human ancestors may well, according to one recent study, have evolved bipedalism (walking on two legs rather than four) in the extensive eastern African forests promoted by the warm Pliocene climate.

This warmth meant that climatically – if not geographically – it was a very different world from today's. A clue to just how different was discovered in 1995 by the geologist Jane Francis – one of only four women ever to be awarded the Polar Medal – in the bitterly cold Transantarctic Mountains. Working with an Australian colleague, Robert Hill, Francis took a closer look at an outcrop of sediments which she knew to be of Pliocene age, and was surprised to discover fossil wood and leaves from beech trees preserved in the rocks.

These fossil leaves didn't come from lush, tall forests like those in England: growth patterns in the wood showed that they came from stunted shrubs which grew prostrate along the ground, presumably because of the harsh climate and strong winds. Even so, the Antarctic of today supports no plants anywhere outside the northern tip of the Peninsula, let alone near the frigid centre of the continent. The fossil site, just above the Beardmore Glacier in the Dominion Range of the Transantarctic Mountains, is a mere

500 kilometres from the South Pole and today enjoys a climate with the average temperature a balmy –39°C.

At the other end of the world, in the Arctic, a similarly surprising discovery was waiting to be made. This time it was northern Greenland – the closest bit of dry land to the North Pole – which yielded ancient Pliocene wood when visited by the Danish geologist Ole Bennike in 1997. Bennike identified some of the pieces as originating from pines and other conifers, in an area which today is hundreds of kilometres north of the tree line. Indeed, the region is now so cold that most of it is completely barren, with just a few areas of herby tundra clinging on in sheltered spots which are watered in the short summer growing season by melting snow. During the Pliocene, clearly, both the Antarctic and the Arctic were much warmer than they are today.

Just how much warmer is indicated by plant and insect remains that have been recovered from an early Pliocene peat deposit on Ellesmere Island, in the far north of Canada. The peat comes from a beaver pond, and the types of plant and beetle remains it contains show a climate with winters 15°C warmer than today. A fascinating variety of mammals were unfortunate enough to end up in the pond: these include a bear, a shrew, a wolverine and even a small horse – all extinct ancestors of today's fauna. The mild climate and rich ecology suggest grassy woodland of larch and birch – all 2,000 kilometres north of the present tree line. Indeed, continental glaciers were entirely absent from the northern hemisphere, contributing to a sea level 25 metres higher than today's.

For many years an argument raged in academic circles about why the Pliocene poles were so much warmer than now. One side maintained that a particularly intense ocean circulation was moving more warm water from the tropics to the polar regions – perhaps because of the closing of the gap between North and South America – leading to a planet with more equally distributed oceanic heat. Opponents suggested that higher carbon dioxide levels in the atmosphere warmed the whole planet more or less equally.

The argument wasn't resolved until 2005 when a massive super-computer modelling study used chemical analyses of seabed sediments to make a sophisticated reconstruction of sea temperatures during the Pliocene. For the ocean circulation crowd to win, the model would need to show cooler tropics and warmer poles. It didn't. Instead, both the tropics and the poles warmed – a finding consistent with carbon dioxide as a cause. As the project's lead author, Dr Alan Haywood, of the British Antarctic Survey, said: 'The sea temperature pattern we found points the finger squarely at CO_2 rather than the ocean currents.' And Haywood drew the obvious conclusion: 'Our findings are critical to understanding how climate may respond to emissions of greenhouse gases in the future.'

The simulated Pliocene has some clear warning signs for today. According to Dr Haywood's model, both the Arctic Ocean and the seas around Antarctica became seasonally ice-free. Winter ice was also vastly reduced. Since ice at the poles acts like a giant mirror reflecting away the Sun's heat, its disappearance works together with carbon dioxide to keep the planet warmer. Indeed, the Pliocene was probably the last time in the past 3 million years that the North Pole lost its ice sheet completely. Ocean circulation patterns also changed. The Atlantic circulation (discussed in chapter 1) was likely also reduced in strength.

But how high were Pliocene CO_2 levels, and how much warmer was it globally? A possible answer to the first question again comes from fossils like the beech leaves Jane Francis found in the frigid Antarctic interior. All plant leaves have tiny holes in them, called stomata, which allow carbon dioxide to enter and oxygen to escape during photosynthesis. Experiments in CO_2-enriched greenhouses show that the number of stomata per leaf changes according to the amount of carbon dioxide in the atmosphere. Many fossilised leaves are so well preserved that their stomata can be counted, and studies of stomatal density in the Pliocene come to a very startling conclusion: atmospheric concentrations of CO_2 ranged from 360 to 400 parts per million (ppm). (This ppm measure means, very simply, that for every million litres of air there were 360 or so litres of carbon dioxide. Expressed as a percentage, the Pliocene atmosphere was therefore 0.036 per cent CO_2.)

If you are reasonably well versed in the global warming issue, the Pliocene CO_2 concentrations might sound rather familiar: today's concentrations are at 382 ppm, and rising at 2 ppm annually. In an article for *Geology Today*, Dr Haywood explains: 'The concentration of CO_2 in the Pliocene atmosphere appears to have been almost identical to atmospheric concentrations today that are a result of the emission of greenhouse gases into the atmosphere.' Temperatures during the epoch have also been pieced together using 'proxy data' from marine and terrestrial fossils dating from the time. So how much warmer was it globally? You guessed it: just under three degrees.

This suggests an obvious conclusion: if CO_2 concentrations at today's levels back in the Pliocene gave three degrees of global warming, surely they'll do the same now? Perhaps – but over a longer time period than a mere century. It takes thousands of years for warmer temperatures to penetrate into the darkest depths of the oceans, for example; and for as long as the seas carry on warming, the atmosphere cannot reach equilibrium, because heat is still transferring downwards. This is an example of the planet's 'thermal inertia': temperatures will always lag behind changes in 'forcing' from solar radiation or greenhouse gases, because of the long response time of the Earth system. In the same way, it takes several minutes for water to boil in a saucepan once the stove has been switched on.

Global warming this century is a result of accumulated greenhouse gases emitted since the dawn of the Industrial Revolution. (It seems amazing to think that coal burned in early steam trains like Stephenson's Rocket is still warming the planet today.) Even if we stabilised atmospheric CO_2 concentrations immediately, it would take many centuries for the Earth to once again reach thermal equilibrium in a new, hotter state. Expecting today's Pliocene CO_2 levels to equate to Pliocene temperatures tomorrow would be like expecting a kettle to boil instantly.

On the positive side, this suggests that if we switch the CO_2 'kettle' off quickly, we can probably avoid hitting three degrees for another century at least. On the other hand, if emissions go on rising as they currently are, global temperatures could shoot past three degrees as early as 2050. The choice is ours, and the clock is ticking.

The Christ Child returns

A barrage of storms pummels the North Atlantic. Floods on the Yangtze River deluge Shanghai, drowning 100,000 people in the rice-growing provinces of Henan, Hubei and Anhui. A great drought parches Amazonia. Crops fail in Australia, whilst famine sweeps northern Africa and India.

The year is not 2050, and the culprit is not global warming. The year is 1912, and the culprit is El Niño. More completely termed the El Niño Southern Oscillation (ENSO) by scientists, El Niño is the warm phase of an ocean current oscillation in the Pacific which can play havoc with the world's weather. The name (meaning Christ Child in its capitalised Spanish form) was first applied to the warm current by Peruvian fishermen who noticed that it tended to arrive around Christmas, decimating their usually productive cold-water fisheries. There is nothing new about El Niño's appearance or the destruction it can cause: ENSO-related droughts over a thousand years ago have been blamed for the collapse of the Moche civilisation in the Peruvian coastal desert. In 1912 it may even have contributed to the sinking of the *Titanic*: unusual storms linked with El Niño drove icebergs much further south than usual, into the path of international shipping, just at the moment the *Titanic*'s Captain Smith ordered his ill-fated ship to proceed at full speed ahead.

Because El Niño effectively reverses the Pacific's weather patterns – causing floods in Peru's Atacama Desert and droughts in Indonesia and Australia – it has knock-on effects (known as teleconnections) right across the planet. On the good side, the north-eastern US tends to experience mild winters, and increases in wind shear over the tropical Atlantic dampen down the Caribbean hurricane season. On the other hand, droughts in forested areas from Amazonia to Papua New Guinea spark devastating fires, whilst rainfall deficits also cause harvest failures and famines in southern Africa. In the nineteenth century El Niño helped trigger the drought which killed millions in British India – indeed a 130-year historical rainfall record shows that monsoon failures leading to severe droughts in India have always been accompanied by El Niño events.

Teasing out the possible linkages between global warming and El Niño has long been a major challenge for climate scientists. The last twenty years have seen stronger and more frequent El Niños, with the 1997–8 event the most powerful on record. Recent evidence from both modelling studies and investigations of past Earth climate suggest that El Niño may not just become stronger, it may become permanent – spelling disaster for human populations and ecosystems around the globe. At the moment transitory floods and droughts mean terrible hardship for the people affected, but sooner or later normality always returns. This may not be the case in our globally warmed future.

However, the scientific community is far from united on the El Niño issue. Some studies do indeed show a shift towards a near-permanent El Niño state in the future, with warmer temperatures in the eastern Pacific and rain-bearing clouds shifting closer to the Peruvian coast. One recent study, however, draws exactly the opposite conclusion, predicting weaker El Niño events in a warmer world. Other work suggests little change, or concludes that the complex computer models built by scientists to replicate the world's climate aren't yet good enough at reproducing the specifics of the ENSO cycle to be reliable predictors of the future. In the modelling world at least, the jury still seems to be out.

But, once again, the past provides useful evidence to help predict the future. Studies suggest that El Niños were weaker or absent altogether during colder periods like the deepest part of the last ice age. Moreover, during the Pliocene warm period discussed above – when global temperatures averaged nearly 3°C warmer than now – there are strong indications that permanent El Niño conditions did indeed prevail.

The reasons are relatively simple: a warmer globe led to a warmer ocean surface, reducing the cold upwelling current that emerges today along the western coast of South America. With less temperature contrast between the west and eastern Pacific, the easterly trade winds weakened or collapsed, driving less water to the west and keeping the cold current cut off. In the process, the low-level stratus clouds that are familiar to every coastal Peruvian (in Lima the endless low mist is called *neblina*) would have been drastically reduced, reflecting less sunlight into space by reducing the planet's whiteness. Hotter temperatures would result, and El Niño's grip would be permanent. Instead of being a reversal of ocean currents, it would be normal. Today the trade winds have already begun to slacken due to human influence, suggesting that the process might now be under way once again.

One man who thinks that 'super El Niños' are a serious danger is James Hansen, the NASA climate scientist whose recent warnings about ice sheet collapse were examined in the last chapter. Although Hansen and his colleagues do not suggest that El Niños will necessarily become permanent, they do present a sound theoretical basis for much stronger and more damaging events. Rapid global warming is already heating up the western Pacific, where – if conditions are right – El Niños are born. This is in contrast to the eastern Pacific, where upwelling waters – having been out of contact with the warming atmosphere during decades spent at depth – remain cool. It is this temperature differential which, Hansen suggests, would provide fuel for the fire of 'super El Niños', sparking weather chaos across the globe.

Europe can expect drier winters in such a scenario. Whilst the Atlantic hurricane season would be moderated by increases in wind shear, hindering storm development, massive floods and mudslides could sweep through normally dry areas of California. With the failure of the Indian monsoon, millions of lives would be at risk in the Indian subcontinent. In South America, one of the wettest areas of all – the great Amazonian rainforest basin, east of the Andes – would rapidly become one of the driest. In the process, as we'll see in the next section, the Christ Child's long-lasting return would light the spark for one of the most destructive conflagrations the world has ever seen.

The death of the Amazon

I have in my filing cabinet a well-thumbed and rather tatty printout of a paper that was published in *Nature* back in November

2000. Stapled onto the back of it is a page of scribbles written by me at the time, expressing startled disbelief at its contents. This paper, perhaps more than any other I had read, convinced me that this book needed writing – not just because of what it said, but because of the reception it got. Despite containing one of the most alarming projections ever published in the scientific literature, barely a ripple was caused in media or political circles. There should have been panic on the streets, people shouting from rooftops, statements to parliament and 24-hour news coverage. There were none of these things. The paper, modestly entitled 'Acceleration of global warming due to carbon cycle feedbacks in a coupled climate model', and written by a team from Britain's Hadley Centre, was largely ignored.

I was reminded of it again during the 2005 Amazonian drought, the worst in decades, a time when fresh water had to be flown into villages which normally sit beside massive rivers. Scanning the headlines, it felt as if the Hadley Centre paper's projections were already coming true – half a century too soon. Fires were erupting in areas that had never burned before. Whole stretches of meandering river had dried up and turned into baked mud. Had Amazonia already reached the tipping point? Thankfully, this time the great forest came back from the brink. The drought lifted before the year's end, and rainwater slowly trickled back into the main river's tributaries, bringing relief to the parched trees. The fires were dampened, and the world's most diverse and precious ecosystem was saved – for now.

The Hadley Centre's *Nature* paper should have rung worldwide alarm bells. First, it showed that global warming could begin to generate its own momentum if a previously unforeseen positive feedback – a vicious circle by which warming would release more greenhouse gas, causing more warming and thereby more gas to be released in an unstoppable spiral – came into effect. This, the 'carbon cycle feedback' referred to in the paper's title, would potentially leave human beings as powerless bystanders in a devastating runaway global warming scenario. Second, the Hadley paper revealed that the main furnace of this positive feedback would burn not in the industrial capitals of the world, but in the remote heart of South America, beginning with the near-total collapse of the Amazonian rainforest.

The paper's authors, led by the climate modeller Peter Cox, had reached this terrifying conclusion by taking what now sounds like an obvious step to make their global model more realistic. Whilst previous models had treated rising temperatures as a simple linear process, Cox's team realised that land and ocean systems would not remain static during rapid global warming. They would *themselves* be affected by the changing climate. In the case of the oceans, warmer seas absorb less CO_2, leaving more of it to accumulate in the atmosphere and further intensify global warming. On land, matters would be even worse. Huge amounts of carbon are currently stored in the globe's soils, the half-rotted remains of long-dead vegetation. According to generally accepted estimates, the soil carbon reservoir totals some 1,600 gigatonnes, more than double the entire carbon content of the atmosphere. As soil warms, bacteria speed up their work to break down this stored carbon, releasing it back into the atmosphere as carbon dioxide. Whereas most climate models treat the land surface as inert, Cox's team for the first time included this 'positive feedback' of carbon releases from warming soils and vegetation – and reached an astonishing result.

According to the newly updated model, a three-degree rise in global temperatures – something that could happen as early as 2050 – effectively reverses the carbon cycle. Instead of absorbing CO_2, vegetation and soils start releasing it in massive quantities, as soil bacteria work faster to break down organic matter in a hotter environment, and plant growth goes into reverse. So much carbon pours into the atmosphere that it pumps up atmospheric concentrations by 250 parts per million (ppm) by 2100, giving a further 1.5°C boost to global warming. In other words, the Hadley Centre's team had discovered that carbon cycle positive feedbacks could tip the planet into a runaway global warming spiral by the middle of this century, much earlier than anyone had so far suggested. By 2100 global warming in the Hadley model rose from 4°C to 5.5°C, perilously close to the IPCC's worst-case scenario. That is why my scribbled notes expressed such shock and dismay when I first read the paper back in 2000.

Politicians may not have stirred, but other scientists did sit up and listen. Following the Hadley Centre's lead, American climate modellers at the National Center for Atmospheric Research in Colorado added a carbon cycle component to their model and also found a decrease in the amount of carbon stored in globally warmed soils. A French team repeated the Hadley experiment, again with a different model, and got the same result. Another team, this time based in both the US and Italy, found that atmospheric CO_2 jumped by 90 ppm and global warming increased by 0.6°C when they added carbon cycle feedbacks to their model. These results may differ in their magnitude, but the direction they point in is the same.

Even if all these models are wrong, another significant danger lies beneath the tropical forest lands of Indonesia, Malaysia and Amazonia: peat. For thousands of years, dead vegetation has built up beneath the living forest, in places forming peat layers tens of metres thick. But this peat layer is only kept stable because it is waterlogged: in Indonesia, it was burning peat that contributed most of the 2 billion tonnes of extra carbon which hit the atmosphere during the devastating fire season of 1997–8. Much of it smouldered underground for months, still releasing carbon even once the overland fires had been put out by the returning rains. This is another potentially devastating carbon cycle feedback: if rainfall patterns shift in a globally-warmed future, leaving these flammable mounds of peat tinder-dry over tens of millions of hectares of south-east Asia and Amazonia, then vast amounts of extra carbon will enter the atmosphere, further aggravating global warming. Modelling results examining this potential feedback are inconclusive but worrying: one 2007 study found that 7 out of 11 models investigated predicted a decrease in dry-season rainfall over the Indonesian peatlands, whilst 6 out of 11 predicted a similar decrease over Amazonia.

This latter result illustrates the uncertainty surrounding projections of future rainfall changes: whilst the Hadley Centre's model shows a dramatic drying trend over Amazonia, some models show less of a change, and others even project increasing rainfall. Which of them is right could hardly be more important, for this colossal ecosystem is home to half the world's biodiversity, and accounts for a tenth of the net primary productivity (the photosynthetic output of plants) of the entire planetary biosphere, in an area of just 7 million square kilometres. Fed by Andean snowmelt and seasonal torrential rains (of over 2.5 metres per year in some regions), the Amazon River contains 20 per cent of all the water discharged into the world's oceans – ten times the volume of the Mississippi. The energy released by this huge amount of precipitation plays a major role in the circulation of the whole world's weather. As the science writer Peter Bunyard explains: 'The functioning of the Amazon basin as a hydrological power engine is a critical component of contemporary climate.'

But Amazonia is already under siege, irrespective of global warming. Over half a million square kilometres – an area the size of France – has been deforested, and more is chopped down each year to make way for cattle ranching and soya plantations. The human population encroaching on the forest has increased tenfold in the last half-century, and each new road the Brazilian government forges into pristine areas is quickly surrounded by new 'herringbone' patterns of deforestation. Slash-and-burn agriculture is also a serious threat, as half a million land-hungry peasants converge on Brazil's last great wilderness in search of a better living for themselves and their families. Illegal logging is rampant: when Greenpeace asserted that 80 per cent of logging was conducted illegally, the Brazilian government – instead of issuing furious denials – threw up its hands and agreed.

Even if all this destruction stopped tomorrow, the Hadley Centre's model suggests that the Amazon rainforest would still be doomed unless global warming levels off at two degrees. If the world crosses this crucial tipping point, the model simulates a tidal wave of destruction, beginning in north-eastern Amazonia and advancing steadily south and westwards across the continent. Modelled rainfall declines almost to zero in some areas by 2100. Temperatures soar to Saharan highs, reaching *on average* 38°C. Once the collapse is complete, the interior of the Amazon basin is essentially desert, devoid of any significant vegetation at all. Only a small amount of grassland and savannah persists on the outer-most edges.

Every fire season gives a preview of how this collapse would unfold in reality. Amazonian trees are used to constant humidity and have no resistance to fire. They are very different from trees in other forest ecosystems which are not only adapted to the occasional burn, but need regular fires to keep the forest healthy. In contrast, Amazonian trees, with no evolutionary experience of burning, continue to die long after the flames have passed. Regeneration is slow, allowing sunlight to penetrate the canopy and further dry out the forest floor. Bird and animal species show catastrophic population declines. And rainfall is further suppressed by the clouds of smoke that hang for months over the stricken forest.

When the final conflagration takes place, it will be on a different scale from anything witnessed so far. The closest comparison might be the 1998 fires across Indonesia, which blanketed several countries in choking smog for many months. In Brazil, Venezuela, Colombia, eastern Peru and Bolivia, life will become increasingly difficult as the air becomes an unbreathable mixture of searing hot gases and smoke. The sun will be blotted out by the leaden pall hanging overhead, whilst a grey drizzle of light ash falls from the sky.

From space, satellites might witness gigantic walls of flame marching through the last areas of untouched forest. Thousands of indigenous people – the Yanomami, the Ashaninka, and other tribes who have known this forest as their only home since pre-history – are driven out. Deprived of their livelihoods and culture, unable to make sense of the sudden disappearance of all they have ever known, they will pine for their lost world. For these people, the Earth itself will have vanished. For once the firestorm has passed, white, grey and black piles of ash, surrounded by seared and smoking timber, will be all that remains of the mighty Amazon.

A new, unrecognisable landscape is born. In the deepest parts of the basin, where once the only sound was the howling of monkeys and the rustling of leaves, a moaning wind has arisen. Dust gathers in the lee of burned-out tree stumps. Nearer to the ground, a gentle hissing sound is heard. Sand dunes are rising. The desert has come.

Australia's Ash Wednesday

The desert arrived in Australia a long time ago – as did fire. Australia is the world's driest continent, and in summer bushfires sweep right up to the edges of cities like Sydney, Perth and Adelaide. There is nothing new about Australian bushfires; some species of eucalyptus trees even need to be burnt over before their seed pods will open. Australia's aboriginal peoples used fire as a way to manage their hunting grounds – in a technique nicknamed 'firestick farming' – long before the white invaders arrived.

But this long history does not make Australian bushfires any less dangerous to people: falling embers can sometimes start new blazes many kilometres away from the main burning area, trapping firefighters between two converging walls of flame. Most feared of all are 'crown fires', which not only speed through the treetops faster than people can run, but can also suck all the oxygen out of the air, asphyxiating anyone caught underneath. On Ash Wednesday, 16 February 1983, twelve volunteer fire crew members on the outskirts of Melbourne were caught out by a crown fire that jumped 500 metres in 15 seconds. None survived. In January 1994 Sydney was almost encircled by over 800 separate fires, which rained ash on the central business district and shut out the sun as brown smoke drifted over vast areas.

In the three-degree world, far more of Australia will burn. Fire risk and intensity depends on the convergence of two key factors: drought and heat. Climate change projections show most of Australia getting hotter and drier over the century to come, seriously worsening fire risk. According to a major study published by CSIRO Atmospheric Research, days of high temperatures above 35°C could increase by 100 to 600 per cent (2 to 7 times) in New South Wales by 2070. Truly searing days when the mercury rises beyond 40°C could increase sixfold in interior bush towns.

Although the models are less clear on precipitation changes, drought frequency could triple, according to the CSIRO study. Average rainfall across the state could plummet by 25 per cent. And extreme winds – which can drive small fires into deadly infernos within minutes – are projected to increase in the summer, when fires do the most damage.

Other states will also suffer. The northern half of Victoria could lose up to 40 per cent of its rainfall by 2070, according to another study by CSIRO. South-western Australia also sees a declining rainfall trend. Not surprisingly, health will suffer in such extreme conditions. Between 8,000 and 15,000 elderly people will die annually from heat-related causes in Australian state capital cities, according to the Australian Conservation Foundation. New diseases could also penetrate: by 2070 the zone of potential transmission for the mosquito-borne dengue fever could reach as far south as Sydney.

Other than fire, Australia's great concern will be water. Despite the country's relative dryness compared to other continents, the average Australian household currently manages to flush, drink, wash and hose its way through a staggering 350 litres of water a day. Such profligacy cannot be sustained as rainfall totals drop and rivers dry up. Northern areas such as Darwin and Queensland – which are already well watered by the Australian monsoon – can expect increases in rainfall as the monsoon intensifies in a warming world. But most of the rest of the country – excepting, perhaps, mid-latitudinal Tasmania – will move into a situation of chronic water shortages and agricultural collapse, a process which is already beginning now as the country reels in the grip of a five-year drought. One study of potential future wheat production in South Australia showed that only farms furthest south and nearest to the coast would remain viable in the future climate.

The Murray-Darling River basin, which extends across five states and supplies water to big cities including Canberra, Melbourne and Adelaide, is projected to lose between a quarter and half of its flow by the time global temperatures approach three degrees. Perth, which has already seen a decline in rainfall in recent years, is even more vulnerable. As the storm systems which bring rain are gradually displaced south over the ocean – the predictable result of the world's regular weather belts contracting closer to the poles – they will be replaced by persistent high-pressure zones of dry weather over the land. The result is straightforward: a big drop in annual rainfall, destroying the future agricultural productivity of a region which currently grows half of Australia's wheat.

This combination of fire, heat and drought will make life in Australia increasingly untenable as the world warms. Farming and food production will tip into irreversible decline. Salt water will creep up the stricken river systems, poisoning groundwater supplies. Higher temperatures mean greater evaporation, further drying out vegetation and soils, and leading to huge losses from dwindling reservoirs stored behind dams.

At the very least, these changes mean big disruptions in everyday life for the average Australian, major economic losses and strict rationing of water. At worst, they may lead to population movements out of areas with too little water, and towards Tasmania and the northern tropical region where rainfall remains more reliable. Life may simply not be possible in much of the interior as temperatures reach scorching new highs. And if the taps do run dry across most of Australia in decades to come, people may well have some harsh words regarding the refusal of the 1996–2007 Howard government to countenance cuts in greenhouse gas emissions. But by then, of course, it will be too late.

Appropriately, Canberra, Australia's capital, got a taste of the new fire regime on 18 January 2003. For several months beforehand, eastern Australia had suffered abnormally low rainfall and searing temperatures. The mercury hit 37°C at Canberra Airport on 18 January, just as strong, dry winds blew in from the west, creating extreme fire danger conditions. No one knows how the first spark was lit, but when the wildfire began, its growth was so explosive that meteorologists compared smoke from the resulting inferno with the volcanic plume from the eruption of Mount Pinatubo.

In just ten minutes of burning at the peak of that firestorm, more energy was released than by the Hiroshima atomic bomb. Enormous fire-driven thunderstorm clouds – termed pyro-cumulonimbus – built up over the flames due to the intense convection and heat. No rain fell, but black hail pounded the ground 30 kilometres to the east. An F2-strength tornado touched down just to the west of the city's fringe. Smoke was thrown into the air with such explosive force that it penetrated the stratosphere and began to circulate the globe – cutting off some of the Sun's rays in a small-scale 'nuclear winter'. When calm was restored, four people were dead and five hundred buildings reduced to ashes.

Since politicians had refused to consider the future, the future had paid a visit to the politicians – in their own home town.

Houston, we have a (hurricane) problem

Houston, Texas: 5 August 2045, 9 p.m.: As the evening light fades, an oily swell has begun to rise in the Gulf of Mexico. A few fluffy clouds catch the dying rays of the Sun, but the sky looks almost too tranquil. Only a thickening veil of high-altitude cirrus gives a hint of what is to come.

On the ground, however, all hell has broken loose. Shop owners are frantically boarding up their stores. The price of plywood has quadrupled, whilst the few shops remaining open have sold out of packaged food and bottled water. A million television screens beam out animated satellite sequences of Super-Hurricane Odessa as she churns over the warm Gulf, whilst urgent official voices warn of the highest-ever sea temperatures driving 180-mile-per-hour winds, and the need to evacuate the entire metropolitan area, *now*. People stream onto shuttle buses. Those left behind commandeer any vehicle they can find, and join the surging mass of humanity desperately moving inland – many on foot, like an army advancing up the Northwest Freeway. Ironically, some are refugees from New Orleans, when the city finally gave up the battle against sea and storms and was abandoned over a decade ago.

Down the river at Galveston, the tide is already surging over the breakwater and into the nearest seafront streets. Hardly anyone is there to see it: folk memories still remain of the 1900 Galveston Hurricane – for more than a century the worst natural disaster in US history – and no one wants to wait on a low-lying barrier island to see if history repeats itself.

The first of Odessa's rain bands advances under the cover of darkness, dumping torrential downpours across coastal Texas, from Corpus Christi in the south right up to the border with Louisiana. This storm is enormous, and Houston is right in the middle of its projected track. Towards dawn the wind rises, and the grey morning light reveals a very different scene from that left behind by the previous night's setting sun. Howling winds roar into Galveston Bay, tearing streamers of water from the building storm surge as it sweeps inland with the speed of a tsunami. Suddenly, rain, sea and wind are barely distinguishable, combining into a morass of water and violence.

Still the storm builds. The surge is now moving up the river, the first water pouring around the buildings on the eastern edge of Houston itself. With blinding rain now pounding all of Harris

County for several hours, Houston's long-tamed river, Buffalo Bayou, begins to return to the wild. First to flood are underground car parks and malls. Storm drains suddenly start spouting flood-water. Manhole covers blow off with no warning, releasing fountains of foam five metres into the air. Abandoned vehicles float down the rapidly rising river, together with wind-blown debris washed out of flooded streets.

As the eye of Super-Hurricane Odessa crosses the coast, she is still a Category 6 monster. All of Galveston is underwater once more, battered by immense waves tens of metres high. As the winds from the storm's eyewall slam into Houston, the gleaming towers of the central business district begin to sway ominously. Squalls of water and spray howl up the concrete canyons, whilst far above, glass windows explode with the force of the blast. Commercial Houston, headquarters of America's oil industry, is ransacked. Blizzards of paper belonging to some of the most pow-erful corporations on the planet rise into the hurricane's central vortex, scattering high into the troposphere. The shrieking winds are so powerful they strip leaves and even bark from the trees in the nearby recreation ground. The sound is like a freight train hurtling through the sky above. Wooden buildings crash and topple – only concrete and steel survives.

On television screens around the world, people are watching the destruction with a mounting sense of dread. From the Bay of Bengal to the Philippines, from Taiwan to Australia, forecasters in all tropical cyclone basins are now on high alert. They know that Hurricane Odessa is only the beginning. The era of the super-hurricane has begun.

As the imaginary scenario above suggests, by the time the world approaches three degrees any lingering doubts about the connec-tion between global warming and stronger hurricanes will have been dispelled by the brutal realities of a more energetic atmosphere. With more energy in hotter seas available to power tropical cyclones, hurricane landfalls will devastate vulnerable coastal areas right across the tropics. New Orleans was the first, but many other cities will follow, from Houston to Shanghai. And two scientists in particular will go down in history for having tried to give us advance warning.

Tom Knutson and Bob Tuleya, two unassuming climate modellers based in the US, wrote a paper in a 2004 issue of the *Journal of Climate* warning about increased hurricane intensity in the greenhouse world. Knutson and Tuleya managed to get computers to simulate hurricanes, complete with their characteristic doughnut shape and calm eye, in order to investigate what might happen to these deadly storms in a warmer world.

Having run their computer-simulated hurricanes an impressive 1,300 times in a doubled-CO_2 climate, they drew a worrying conclusion: as global warming intensifies, so will tropical cyclones, with maximum wind speeds rising by 6 per cent, and rainfall by 18 per cent. These figures may sound small, but in the real world they will mean more highly destructive Category 5 storms battering shorelines across the world. In our current climate, Category 5 is the top rank for a hurricane. But in a warmer future, with all storms potentially half a category or more stronger, the worst storms will be more deadly than anything we have ever experienced so far.

Dawn over a new Arctic

As a fully-fledged hurricane demonstrates with near-perfect symmetry, the atmosphere is a fluid in motion, full of waves, swirls and eddies just like a flowing river as it constantly moves over the surface of the Earth. The swirls are low-pressure systems – of which tropical cyclones are one type – whilst the calm pools are highs, where sinking air leads to gentle winds and dry days. Like water in a river, all the atmosphere is interconnected – not just with itself, but with the oceans and the biosphere too. Hurricanes, for example, mix the upper layers of the tropical ocean with their strong winds, helping to drive the massive currents which eventually bring warm water to the poles.

Changes at the poles will therefore have knock-on effects in other areas far away. One likely outcome is that a reduction in Arctic sea ice will exacerbate the drying of western North America. Instead of ocean heat remaining trapped under surface ice during the winter, once most of the ice cap has disappeared large areas of open ocean will remain exposed to the winds, altering the usual pattern of winter weather over North America. In one modelling study, the rain-bearing low-pressure systems get shunted further north towards Canada and southern Alaska, and away from the drought-scarred plains of the western US. The result is a precipitous 30 per cent drop in rainfall throughout the entire west coast of America. Water shortages propagate far inland, bringing drought emergencies to eleven US states from Nevada to Wyoming. Moreover, as discussed in chapter 1, the changes are but one part of the wider tendency of world weather belts to contract towards the poles as global temperatures rise.

In the Arctic itself, model simulations suggest that 80 per cent of the sea ice will have already been lost as the world approaches three degrees, with only a small patch hanging on between the pole and the north coast of Greenland. As we saw in the previous chapter, this projected 80 per cent decline is probably a serious underestimate – more likely is that all the ice will have disappeared. Most models show an Arctic which warms much more slowly than is already happening in the real world: one recent analysis showed that sea ice declines are currently 30 years ahead of those projected in the IPCC's models. The authors, some of the world's leading experts on the Arctic, reasonably conclude that 'the sensitivity of this region may well be more than the models suggest', and that the 'transition to a new Arctic state' will occur 'well within this century'.

On land, ice caps and glaciers will also be melting rapidly. The three-degree world will see water sluicing off Greenland in phenomenal quantities, converging into immense glacial rivers as the ice edge retreats into the centre of the giant island. As the glaciers pull back, new areas of land open up which have been under the grip of ice for hundreds of thousands of years. Huge new lakes form, trapped between dams of gravelly moraine and the retreating ice cap. With the melt zone creeping inland each summer, blue lakes now appear across the whole ice sheet. Another positive feedback swings into operation: at the moment, the summit of Greenland is colder than the base by dint of its sheer altitude (just as the peak of a mountain is cooler than its lower slopes), but as melting increases, less and less of the giant ice cap keeps itself cool by staying high up in the atmosphere. With more of the ice sheet stuck in warmer temperature zones, melting accelerates still further.

Smaller ice caps, such as Iceland's Vatnajökull, will disappear much faster than the lumbering giant of Greenland. Familiar to the tourists who flock to Iceland's volcanoes, geysers and waterfalls, Vatnajökull is a big dome that dominates the south-eastern side of the country. The largest ice mass in Europe, it is underlain by several volcanoes; some of which occasionally erupt underneath the 400-metre-thick ice layer, causing the famous *jökulhlaup* floodwater outbursts every ten years or so. In total Iceland's glaciers contain about 3,500 cubic kilometres of ice – enough for a 40-metre-thick layer of ice across the whole country.

By 2100, only half of Vatnajökull is left, and by 2200 the

whole ice cap has gone. River run-off doubles as the meltdown gets into gear, sending huge amounts of muddy water gushing across the plains and into the sea. The smaller Icelandic ice caps like Hofsjökull disappear even faster. Alpine mountain glaciers in Sweden and Norway also melt rapidly, along with those in Alaska and northern Canada. Although compared to the polar ice masses of Greenland and Antarctica their water content is tiny, if all the small ice caps and glaciers were to melt together, global sea levels would rise by a quarter of a metre.

These changes are reflected in other areas of the Arctic. On the positive side, perhaps, growing seasons get longer – allowing crops to be grown further north than ever before, and perhaps helping to close some of the food gap arising from the drought-affected farmlands further south. By 2050, for instance, Norway enjoys a growing season equivalent to today's in southern England, the Netherlands and northern Germany. In Finland, the growing season increases by up to two months.

It is unlikely, however, that major new areas of farmland can be opened up: most of the Arctic is rocky and acidic, with only thin soils. Some areas of formerly frozen bogland may dry out and be available for ploughing – but this would come at the expense of enormous carbon emissions from the decomposing peat. In a region where the predictable cycles of the seasons will have long since disappeared, this may not be the only surprise lurking just around the corner.

Mysteries of the Maya

Many visitors to Mexico's Mayan ruins feel that there is something special about Palenque. Its pyramids aren't as huge as those at Chichen Itza, and its temples aren't as remote and photogenically mist-wreathed as those at Tikal. Perhaps it is the knowledge that only 5 per cent of the ancient city has so far been excavated, and the exciting sight of mysterious mounds still covered in trees next to tracks leading off into the jungle. Or the recently deciphered hieroglyphs telling of supernatural deeds by holy kings and brave warriors. Or the lushness of the surroundings, where natural springs gurgle from the rocks to join the river Otolum, which the Maya architects channelled symbolically underneath Palenque's main square.

Whatever it is, Palenque's sheer beauty and mystery combine to make the site seem other-worldly and somehow transcendent. As an ignorant student backpacker, I found it difficult to believe that the ancestors of the simple-looking, barefooted indigenous people hawking fruit at the roadside had centuries earlier been responsible for creating such extraordinary monuments. Early archaeologists had the same problem, especially given the sparse population found in the area when the first Europeans visited. How could these villagers in their mud huts have carved such wonders? And why were these great cities abandoned? Many preferred to attribute Palenque and other sites to the Egyptians, the Polynesians or even the Lost Tribes of Israel, especially as the local people had little knowledge or folk memory to explain them.

As more became known about the Maya, the mysteries only deepened. This had clearly been one of the most advanced societies ever to emerge in the New World. Millions of people had lived on irrigated agricultural lands as densely populated as modern-day Bangladesh. The Classic Maya (between 50 BC and AD 900) made extensive use of stone-carved writing, and even manufactured books out of bark and plaster, the mouldering remains of which have been found in royal tombs. They employed a calendar, known as the Long Count, which dated back to 3000 BC and profiled major battles and dynastic successions. They even, it has been suggested, understood astronomy and predicted lunar eclipses.

Yet by the time the Spanish invaded, the Classic Maya cities of Palenque and Tikal had already been abandoned for centuries, their tall pyramids crumbling into the advancing forest. Only a few farmers remained nearby, eking out a living from the poor soils by growing stunted maize and beans. As excavating work continued, it gradually dawned on archaeologists that something dramatic and terrible had happened to the Maya, something which had precipitated the collapse of their entire civilisation almost overnight.

That something, it turned out, was drought. Most of the Mayan areas are underlain by limestone, which retains little water in times of low rainfall. The Maya compensated for this by building plaster-lined reservoirs and aqueducts, and digging massive wells – called *cenotes* – in low-lying northern areas to reach the water table. But the drought clock was always ticking: Mayan reservoirs could only hold enough water for 18 months.

This was of great concern to both kings and peasantry: the cultural glue that held Mayan society together was a kind of compact, whereby in return for luxury and adulation the kings would perform sacrifices and rituals in order to satisfy the gods and keep the rains coming on time. In times of perennial drought, the ruler's claim to divinity would begin to look rather threadbare, and his once loyal subjects might contemplate revolt or even regicide once starvation loomed.

When the long drought struck, it seems, that's exactly what happened. Lake-floor records drilled from the region show clear evidence of long-term dryness at just the time of the Mayan collapse. Further evidence from ocean sediments confirms that the Mayan Classic Period ended with the double blow of an extended dry period 'punctuated by more intense multiyear droughts' between AD 810 and 910.

In his book *Collapse: How Societies Choose to Fail or Survive*,

Jared Diamond identifies the Maya collapse as a classic case of ecological overshoot, where a highly developed society overreaches its resource base, leaving itself vulnerable to a natural disaster like drought. Diamond also warns that population growth and economic development are once again putting the resources of Central America under great pressure. Large areas of southern Mexico have again lost their tree cover. Hillside erosion is again destroying farmland. Flood-related mudslides sometimes kill thousands when a major hurricane makes landfall. And what if drought strikes again?

Drought, unfortunately, is central to forecasts for the region in the three-degree world. Although precipitation in the deep tropics is projected to increase, the subtropics get drier, and Central America is right in the middle of one of these drying zones. The Hadley Centre model predicts rainfall declines of 1–2 mm per day, half the total annual rainfall in some areas. And just as during the Maya collapse, a drier average situation will mean more intense droughts striking more regularly. Dry weather can also worsen deforestation, by killing off the remaining trees. Largely for this reason, Central America was identified in a 2006 study as one of the world's climate change 'hot spots', areas which are of the greatest concern on the planet.

It is hard to see how the tens of millions who currently live in Central America could survive for long in such marginal lands. A foretaste was given by a moderate drought in 2001, which triggered food shortages amongst an estimated 1.5 million people, leaving hundreds of thousands dependent on food aid for several months. Such aid will not materialise in a world where food supplies are already stretched to breaking point. So as climate change accelerates, these vulnerable Central American countries will be among the first to see their agricultural productivity crippled and their people marginalised and displaced. As whole areas are depopulated, emigrants will leave behind entire ghost towns, little more than dusty ruins in lands which can no longer support them.

These abandoned villages and towns may lie untouched for centuries, just as the ruins of their residents' Mayan ancestors did before them. They too will hold a lesson for a future world. But whether any humans will be around to learn it is far from clear, as the following chapters will show.

Mumbai's monsoon

The crops of 60 per cent of the world's population depend on a single recurrent weather feature: the Asian summer monsoon. Millions go hungry in years when the monsoon fails to bring its promised rains, and millions more lose their homes to flooding when the monsoon brings more of a deluge than the land can hold. Daily totals of 40 cm of rain are not uncommon during these intense tropical downpours, and in Mumbai's record-breaking cloudburst in July 2005 nearly a metre of rain fell in less than 24 hours – the resulting floods causing 1,000 deaths. But without this annual watering, the fertile plains of India and Bangladesh – which support some of the world's densest populations – would be arid and lifeless. The reliability of the monsoon in a warmer future is therefore a matter of life and death for millions of people.

As I explained in chapter 1, monsoons depend on heat differentials between land and the surrounding ocean. These land–ocean contrasts will intensify in a warmer world, meaning that a total collapse of the Indian summer monsoon is unlikely over the next century. Indeed, times in the past when the monsoon has been weaker tend to be associated with colder periods, such as the height of the last ice age. Modern-day pollutants like aerosols (mostly sulphur dioxide and particulates released by fires) can also have a cooling effect, and in the absence of global warming the so-called 'Asian Brown Cloud' might severely disrupt the monsoon, according to one study. The authors hypothesise that a 'roller-coaster scenario' might still come into play, however, where cooling pollutants first weaken the monsoon, before pollution controls and increasing greenhouse gas levels bring it back with greater force than ever. But many different studies – looking at the impact of greenhouse gases alone – have reached the firm conclusion that heavier monsoon rainfall is the most likely outcome, especially given the fact that a warmer atmosphere holds more water vapour.

An intensification of the monsoon presents two problems. First, an increase in average rainfall during the monsoon will mean more heavy events and associated extreme flooding, thanks to this warmer, wetter atmosphere. An increase in the intensity and frequency of heavy rainfall events has already been identified over India, as I explained in chapter 1. Second, the monsoon is also likely to become more variable. Some years it may fail altogether in some areas, only to return with renewed vigour the following season. Forecasters already struggle to predict the monsoon's annual ebb and flow, and this job will become even more difficult as the world's weather shifts into new patterns.

These new weather patterns may affect areas a surprisingly long distance away from the Indian subcontinent. Studies of old corals off the coast of Indonesia have shown that 6,000 years ago, during an era of strong monsoons, Sumatra and other Indonesian islands were struck by harsh drought, thanks to changing wind-flow patterns across the tropical Indian Ocean. This suggests that drought may once again become perennial in this densely populated country if the Indian monsoon gets stronger, as many climate models predict. Drought years in Indonesia can already be disastrous, with food shortages and forest fires so severe that huge areas of the country – and neighbouring Malaysia – disappear underneath heavy palls of smoke for weeks at a time. Areas even as far away as southern Australia may also suffer reduced rainfall as a result of the changing Indian monsoon.

In India itself, farmers will struggle to grow successful crops if the monsoon's onset becomes less reliable. In more intense years, flooding will become catastrophic, submerging vast areas. Particularly badly affected will be the areas that already get the heaviest rains – the Indian west coast, the Bay of Bengal and Bangladesh, and north-east India. It is almost like a biblical adage: to them that already have, more will be given. And the reverse is also true: rainfall may reduce in areas which currently receive scanty amounts, such as northern India and southern Pakistan.

Indeed, it is to Pakistan that we look next, because as world temperatures climb towards three degrees, this country will be on the brink of a crisis without parallel in recent human history. And as is so often the case, this disaster will flow not from the direct effects of the increased heat, but from its indirect consequences. Pakistan stands to lose massive quantities of the most precious resource of all: water.

Where the Indus once ran

The world's two highest mountains both straddle international borders. Everest forms the boundary between Nepal and Tibet, whilst K2 divides Pakistan from China's remote Xinjiang province. K2 also forms the apex of the mighty Karakoram range, which boasts four peaks over 8,000 metres, and another ten over 7,000. One of the world's highest and most inaccessible mountain ranges, the Karakoram includes the largest glaciated area outside the poles. Climbers must walk for a week up the Baltoro Glacier, for example, before they even get their first view of K2's forbidding pyramidal peak.

These glaciers represent a massive store of fresh water, and all three main rivers on the Indian subcontinent – the Indus, Ganges and Brahmaputra – rise amongst the snows of the Karakoram, Himalayas and the great Tibetan plateau beyond. Since the Mekong, Yangtze and Yellow rivers also have their sources in this gigantic upland area, it is no exaggeration to say that rivers rising on the slopes of these glaciated peaks currently sustain half the world's population.

Whilst waters from the Everest region eventually enter the Ganges, run-off from the north side of the Himalayan continental divide joins the Brahmaputra, which runs from west to east before performing a U-turn through the mountain chain east of Bhutan and south into Bangladesh. The Indus, meanwhile, derives much of its early flow from the 72-kilometre-long Siachen Glacier, one of eight glaciers in the Karakoram region longer than 50 kilometres. Further downstream, the Indus runs right past the towering 8,126-metre peak of Nanga Parbat, at the far western end of the Himalayas, where another cluster of glaciers adds further run-off. At this point the Indus valley is already essentially arid, and views of the towering slopes of this giant mountain – ninth highest in the world – are a relief from the shimmering heat haze along the flat and stony valley floor.

All told, the Indus basin includes more than three and a half thousand individual glaciers – more than both the Ganges and the Brahmaputra put together. Right along its length, however, the Indus experiences a more arid climate than the two other rivers, both of which gather more run-off from the summer monsoon rains in their lower reaches. This combination of aridity and a highly glaciated catchment makes the Indus particularly dependent on mountain water. And this is its Achilles heel: the once timeless mountain snows are already melting. A quirk of atmospheric dynamics means that air temperatures in the highland region are rising at twice the global average, and accelerated glacial retreat has been recorded across all the major ice-capped ranges. Even mighty Everest has been losing glacial mass: the Khumbu Glacier has retreated over 5 kilometres from the spot where Hillary and Tenzing set out to conquer the peak back in 1953.

As the glaciers at its source gradually recede over the century ahead, the Indus will change dramatically. In the early stages of this great transformation, the river will be swollen with extra melt as the glaciers continue their accelerated retreat. For a developing country with expanding agriculture and population, the increased fresh water supply may seem like a boon for Pakistan – but it will be a short-lived one. In later decades, once the majority of the ice has gone, the river's flow will slow to little more than a trickle, and the great, braided plain of the Indus River will lie empty for months at a time.

A study funded by Britain's Department for International Development, which used hydrological modelling to project changes in river flows over the century, confirms the grim forecast for the upper Indus. Whilst flows initially increase by between 14 and 90 per cent depending on the speed of the warming, later in the century the model projects decreasing water supplies by up to 90 per cent. At Skardu, gateway to the Karakoram and a major stopping point for high-altitude climbers heading to K2 and Broad Peak, modelled flows on the Indus River peak in 2030 but then drop by half by 2080, whilst further upstream still, at Shyok in Ladakh, the river could be running dry by 2090. Downstream, meanwhile, where the Indus begins to leave the mountains at Bisham Qila, flows will have diminished by between 20 and 40 per cent by 2080.

The disappearance of glacial meltwaters from the sources of the Indus will have a dramatic effect on downstream flows right the way through Pakistan. Little rain falls either in the east, on the Punjabi plains, or in the west, in the dry and lawless mountains of Balochistan, to augment the glacial flow. With Pakistani agriculture almost entirely reliant on water from the Indus flowing into the irrigation canals which criss-cross the Punjab, the country faces the destruction of its key breadbasket once the river and its tributaries run dry.

Little help will be available abroad, as countries which border Pakistan are also likely to experience problems with declining water supplies. In China, for example, 23 per cent of the population lives in western regions where glacial melt provides the principal dry-season water source. Even though it will be less affected than the Indus, as much as 70 per cent of the dry-season flow of the Ganges, and half the flow of other major rivers, comes from melting glaciers. In some ways, these are 'fossil' water supplies, which have been stored for millennia when the glaciers were in relative balance, but will be released in decades to come as the world warms up.

With India particularly dependent on hydroelectric power generation, dwindling summer flows may lead to blackouts and energy shortages during the hottest months of the year. Two of the Indus River's major tributaries – the Chenab and the Sutlej – arise in India and flow into Pakistan. Both will also be suffering the effects of deglaciation in their upper reaches. Conflicts may well break out between these two nuclear-armed countries as water supplies dwindle and political leaders quarrel over how much can be stored behind dams in upstream reservoirs.

Any crisis in food production could quickly escalate into a crisis for the whole Pakistani economy. The country depends heavily on cash-crop exports of rice and sugar, both of which are grown extensively in irrigated Punjab. Further south, where the province of Sindh relies solely on the Indus for inflows into its canal system, farmers may find themselves outcompeted by their more powerful Punjabi colleagues to the north. Farmers across the country may find themselves outcompeted in turn by big cities like Lahore, Hyderabad and Karachi, which each support populations in the many millions.

As dry-season supplies from the Indus dwindle with the retreating ice upstream, a cascade of impacts may begin to pile up the pressure on Pakistani society, as people leave the land in huge numbers and flee to the overcrowded cities. With global food stocks already under pressure, as discussed earlier, little or no surplus capacity will remain to feed people displaced from formerly irrigated areas. Further conflicts could arise with India if millions of Pakistani refugees cross the border in order to find sustenance in better-watered areas served by the Ganges but already supporting dense human populations.

All of human history shows that given the choice between starving in situ and moving, people move. In the latter part of the century tens of millions of Pakistani citizens may be facing this choice, as the river which has sustained their civilisation for centuries runs dry, and the breadbaskets which it supported are overcome by the spreading desert. Nuclear-armed Pakistan may find itself joining the growing list of failed states, as civil administration collapses and armed gangs seize what little food is left. The rule of the gun would replace the rule of law.

In the high mountains, meanwhile, the shimmering white of all but the highest snows will have given way to bare rock and sun-baked soil. Down below, valley glaciers which for millennia ground their way slowly between the peaks will have vanished into rubble. And somewhere in the remote wastes of the Tibetan plateau, far away from any human habitation, a simple pile of stones will mark the spot which was once the source of one of the mightiest rivers in history.

The last drops of the Colorado

Unlike the Indus, the Colorado is no longer a natural river. Its ebb and flow is controlled not by the vagaries of the weather but by engineers pressing buttons in the control rooms of over a dozen major dams. Indeed, every last drop of the 18 billion cubic metres of water that travel down the Colorado River in an average year is already allocated to human use. Water is apportioned between the Upper Basin US states (Wyoming, Utah, Colorado and New Mexico), Lower Basin US states (California, Arizona and Nevada) and Mexico by a raft of agreements. Like the Indus, most – over 70 per cent – of its flow comes from its high-altitude upper basin in the mountains, rather than from its arid lower reaches. But like California's Sacramento and San Joaquin rivers, the Colorado basin is less a living river and more a giant plumbing system.

Moreover, the Colorado's plumbing is already creaking with overuse. A series of drought years in the early 2000s nearly pushed the system over the edge, with water levels in Lake Mead – behind the narrow Hoover Dam – falling low enough to endanger supplies to three states and millions of people. Although the Colorado dams, which can hold four years' worth of river flow, act as a buffer to changing river levels, long-term drought is still a critical threat. One detailed modelling study of how the Colorado system might behave in a future climate found that for between a quarter and half the time during the second half of the twenty-first century, the system would essentially fail.

The Colorado dams have generated major opposition ever since they were first mooted, particularly the Glen Canyon Dam, which drowned beautiful sculpted canyon walls – the famous

Cathedral in the Desert – in an area which has since been called America's 'lost national park'. Campaigners assert that so much water is lost to evaporation that Lake Powell, behind the dam, is essentially useless, and storage could better be conducted elsewhere.

Whatever the rights and wrongs of this particular case, it is clear that some way of storing river water is essential to balance out seasonal changes in flows. This is particularly the case with rivers in the US west, which reach a peak with the early summer snowmelt and then decline later in the summer when human demand is highest. The problem of declining snowpack was discussed in the previous chapter, where we saw that west coast states face big changes in river run-off and water supplies to cities and agriculture as the world warms. By the time average global temperatures rise towards three degrees above today's, the situation will be increasingly critical. Not only will snow have turned to rain in lower-elevation basins, meaning no natural winter storage at all, but the average date of spring snowmelt from the higher mountains will be coming a month or more earlier in the year.

With more winter rainfall and earlier spring snowmelt, water will need to be released from dams to leave spare capacity to safeguard against flooding, wasting a precious resource which will become scarce later in the year. Most river systems, however, do not have enough dams to control much of this run-off, so the majority of this 'early water' will be passed on to the oceans and lost. The long summer drought will become longer still, threatening ecosystems and human habitation throughout the whole of the US west.

All of the Pacific coast of the United States, in fact, is critically dependent on snowmelt. Seattle, for example, already experiences shortages in low-snowpack years. Agriculture will also be affected. Washington state's fertile Yakima Valley, famous for apples, cherries and a thriving dairy industry, gets precious little rainfall – and its irrigation systems depend almost entirely on vanishing snowmelt.

Even far to the north in western Canada, where one might imagine that the winter cold would protect the seasonal snows, this situation will be repeated. One Canadian river stands to lose 40 per cent of its water. The Rockies could become virtually snowless across vast areas, with only the highest peaks retaining a winter covering. Irrigation water from the Rockies is vital to farming throughout Alberta, and crops, like potatoes, pulses and spring wheat, will quickly fail without it.

All this is without considering an additional danger: drought. Chapter 1 showed how much of the western United States is vulnerable to a return of medieval mega-droughts in a warmer world, and the south-west in particular – California, Arizona, Nevada, Utah, Colorado, New Mexico, Texas, Oklahoma and Kansas – can expect a permanently drier climate regime, according to an unusually strong consensus of climate models. According to a May 2007 paper published in *Science* magazine, this increasing level of aridity would be similar to the Dust Bowl years or the 1950s droughts – except that this time the change will be permanent.

Snowless springs, hotter summers, and harsher droughts will all increase the vulnerability of the US south-west to perhaps the most feared agent of all: fire. California has historically been particularly prone to wildfires – it has suffered more financial losses than any other state over the past century. Already there is evidence that the number of severe wildfires suddenly increased in the mid-1980s, thanks to longer summers and earlier snowmelt – and the fires have been burning for weeks at a time. In 2003 wind-whipped wildfires tore across southern California, even reaching the outskirts of Los Angeles. Firefighters could only stand and watch as massive tornadoes of flame rose above the conflagration.

According to projections for the future drawn up by the USDA Forest Service and academics based at Lawrence Berkeley National Laboratory, the number of 'escaped' out-of-control wildfires could increase by over 50 per cent in the San Francisco South Bay area, and soar by 125 per cent on the slopes of the Sierra Nevada, further east. Only the foggy and damp northern coast would escape the flames. Nor would California be the only affected area: most of the Southwest, the Great Basin and the northern Rockies would face two to three weeks more of high fire danger each year.

All these different impacts point to a single outcome: a very different US west from the one we know now. As the mountains lose their snow, so the cities lose their water and the farmers lose their fertile fields. As summer droughts dry out grassland and forest, people will wait for that single spark which will set their world aflame. Fire crews may arrive, but their trucks will be empty and their hoses useless. There will be nothing to stop the burning.

Sinking the Big Apple

On the other side of the American continent, New York City will be facing the opposite problem. Instead of a shortage of water, the great threat to the Big Apple is too much of it. The New York metropolitan area, home to nearly 20 million people, has 2,400 kilometres of coastline – most of it low-lying and heavily built up with apartment blocks, roads and rail links. Four out of five of the city's boroughs are located on islands. More than 2,000 bridges and tunnels connect these islands and the mainland, and most rail, tunnel and airport entrances lie at elevations of only 3 metres or less. The configuration of the east coast shoreline also makes New York particularly vulnerable to storm surges – the right-angle bend between New Jersey and Long Island funnels water right into the city's harbour, whilst the eastern end of Long Island forms another surge funnel.

As city officials are fond of saying, when it comes to a great flood, the matter is not if but when. Had Hurricane Floyd not weakened into a tropical storm when it passed over the area in 1999, the 'big one' might have already happened. As it was, 30 centimetres of rain fell in some areas, causing flash flooding and nearly shutting down the metropolitan transportation system entirely.

A few years earlier, in December 1992, the city came even closer to inundation: a strong nor'easter pushed a massive storm surge onshore at the same time as pelting the whole region with heavy rain. The surge shorted out the whole New York City subway system, stranding people on trains and in stations. It also submerged FDR Drive and other roadways in Manhattan under more than a metre of water, whilst two metres flooded into Battery Park tunnel. Seaside communities were evacuated in New Jersey, Connecticut and Long Island. Officials breathed a sigh of relief when the storm began to move away: they knew that flood levels just half a metre higher would have resulted in massive inundation and loss of life.

Sea levels along the New York coast have already risen by 25 cm over the last century, and this rate is projected to accelerate dramatically over the next 100 years, with the global component of the rise aggravated by the gradual sinking of the land along most of the US east coast. By the time world temperatures approach three degrees above today's levels, sea levels will have risen by anything between another 25 cm and a full metre (much of the reason for the uncertainty, as we saw earlier, is the unpredictable behaviour of the Greenland and Antarctic ice caps). With stronger hurricanes likely, and possibly stronger winter nor'easters too, the New York 'perfect storm' may happen not once but many times over. What counts today as a one-in-100-year flood could arrive every 20 years by the 2050s, and every 4 years by the 2080s. The 3-metre flood zone, which today includes much of Lower Manhattan, Coney Island in Brooklyn, substantial parts of Jersey City and Hoboken as well as both Newark and LaGuardia airports, could be submerging as frequently as every five years, making whole zones economically unviable. Flooding problems will be worsened by coastal erosion, as rising sea levels gradually eat away at the shoreline. Without massive sand replacement exercises, beaches in northern New Jersey and Long Island could be moving inland by up to 3 metres a year by the 2080s.

Already various planners have mooted the possibility of constructing flood barriers at three locations to protect New York from future storm surges: between Staten Island and New Jersey at Arthur Kill, the Narrows at the mouth of New York Harbor, and across the upper East River at the north shore of Queens. 'In future, a weaker storm will do the same damage that a severe storm does today,' says Malcolm Bowman, a surge expert and flood barrier advocate. 'These three barriers, high enough to withstand any conceivable storm surge, closed before a major storm, would provide a ring of shelter around the city.'

Constructing such a series of barriers would cost billions of dollars, but *not* constructing them might cost even more in terms of losses to life and property. Three of the most vulnerable parts of New York – Lower Manhattan, Coney Island and Rockaway Beach – are also densely populated, and evacuation routes will lie below storm-surge flood levels, cutting people off from lines of safety. As the decision makers who did nothing about the levees around New Orleans found out to many people's cost, by the time the big storm approaches it is already too late.

Storms gather in Europe

It did indeed take billions to construct the Thames flood barrier, which protects London from storm-surge floods, but it was money well spent. The barrier was raised 62 times between 1983 and 2001 – and with increasing frequency in later years. Its successor (the current barrier's design life ends by 2030) may need to be raised as many as 200 times every single year by the end of this century to cope with the combined impact of stronger storms and sea level rise.

Jason Lowe at the UK Met Office's Hadley Centre is one of several scientists working on the problem of predicting future storm-surge danger. He co-wrote a paper in 2001 which suggested that flooding events around the British coast might indeed increase in a warmer world. 'In the southern North Sea,' he says, 'by the 2080s, a typical return period for what is now a 150-year event will be seven or eight years.'

A storm surge was responsible for the UK's worst-ever natural disaster in 1953, when more than 300 people died as a howling North Sea storm flooded 24,000 homes along England's east coast on the night of 31 January. Many survivors spent hours huddled on their roofs in the freezing darkness as waves washed through their lower floors. Only the lucky ones made it through the night – many more succumbed to the bitter winds and hypothermia before rescuers could reach them by boat the following morning.

At the time, the 1953 calamity was classified as a once-in-120-year event. Lowe's work shows that flooding of this kind could hit the coast every few years by the latter part of this century, leaving whole villages, towns and huge swathes of farmland uninhabitable. Cities, with their expensive real estate, can be protected. But for the coastline as a whole, retreat inland is the only viable option as the waters rise.

The frequency of storm-surge events depends on the frequency of the extreme weather that generates them. One study by Germany-based researchers projected that by the time world temperatures approach three degrees, more extreme cyclones will track across western Europe, with more storm-wind events striking the UK, Spain, France and Germany. A second study foresees intense cyclones becoming more frequent worldwide by the second half of the century, even whilst the overall number of storms decreases. Another study projects stronger storms particularly across western Europe north of the Alps – encompassing in particular France, Germany, Denmark and the UK – with 25 per cent stronger winds in prospect. As a result, more gales will sweep across the North Sea, leading to stronger storm surges on the coasts of Holland, Germany and Denmark.

With its low-lying coast and extensive array of barrages and dykes, the Netherlands will be particularly vulnerable to sudden changes in sea levels beyond those envisaged by the engineers who constructed this coastal protection infrastructure, none of whom would ever have heard of global warming. Many more Dutch died in the 1953 disaster than English – the final death toll was over 1,800 people – and with large areas below current-day sea levels, Holland may have to accept a sizeable contraction in its geographical area by the middle of this century. Given that much of the Netherlands lay underwater before the reclamation activities of the Middle Ages, the country's current enlarged shape could prove to be a temporary aberration in the ongoing ebb and flow of the North Sea.

The increasing wind will be accompanied by heavier rain – something projected for the whole globe as more heat energy speeds up evaporation and precipitation. Indeed, the process has already begun: a planetwide intensification of the hydrological cycle has already been observed. In July 2007 an international scientific team reported in the journal *Nature* that regions in the northern hemisphere poleward of 50°N (an area that includes Canada, the UK, Scandinavia, northern Europe and Russia) have experienced an increase in precipitation, this change roughly balanced by a drying in the tropics and subtropics. This trend is projected to continue: northern Europe stands to experience 20 per cent more rainfall by the 2070s, with most of this coming during increasingly intense heavier events. This means more winter floods: the north and west of the UK could experience a 50 per cent increase in peak flooding, according to one study.

Droughts and floods could come in alternating years, just as the devastating European deluge of summer 2002 was succeeded by a lethal heatwave in summer 2003. Indeed, a study by Italian climatologists projected an increase in both severe floods *and* severe droughts for western and central Europe in a warmer climate – leaving agriculture and population centres battered by deluges and water shortages in quick succession. With sharply reduced snowfall in the Alps, Europe will suffer similar problems to the western United States during dry years, when rivers peak early in the spring and then decline to dangerously low levels in the high summer, stranding barge traffic and leaving crops to wither in the fields. The Rhine, western Europe's longest river, could be seeing 30 per cent more flow in winter months – causing repeated flooding downstream in Germany and the Netherlands – whilst flows diminish by 50 per cent in August.

South of the mountains, parts of the Mediterranean rim will be transforming towards desert conditions: French researchers, for example, were surprised when their model showed a decrease in summer evaporation from Mediterranean soils despite soaring temperatures by the 2070s. This was odd because with higher temperatures water evaporates faster, so summer evaporation should have gone up. It was only after a thorough investigation that they realised that this was because the soils would be so parched that there would be no water left to evaporate. Nothing could grow under such conditions. The Sahara will have jumped the Strait of Gibraltar and begun marching north.

Africa's fever

It is not just Europe, of course, that will experience a wetter climate and more extremes in a warmer world. East Africa also stands to get more humid, even at the same time as the sand dunes pile up in Botswana and the south. East African rainfall is complex: the region sees two wet seasons – one in spring, and a second in autumn – due to the seasonal shifting of the trade winds and the intertropical convergence zone (the belt of thunderstorms that girdles the Earth's equator and moves with the summer into each hemisphere). Countries from Somalia down to Mozambique, including Kenya and Tanzania, could see more extreme rainfall and frequent floods. In Europe pools of stagnant water left by floods may be little more than a nuisance, but in Africa they bring with them a deadly companion: disease.

Malaria and other vector-borne diseases like dengue fever thrive in warm, wet conditions, and climate change is expected to help both diseases move to higher altitudes and latitudes as temperatures and rainfall rates rise. Africa is ground zero for malaria: 85 per cent of infections and deaths occur on the continent. As malarial mosquitoes migrate uphill into newly suitable areas with the warming climate, so whole new populations – many of whom have never experienced malaria before – become at risk. Zimbabwe could be particularly hard hit. Most of the population currently lives above the malarial transmission zone, because major cities like Harare and Bulawayo are located high up on the country's central plateau. But as temperatures rise, the *Anopheles* mosquito that carries malaria will be able to fly unimpeded across the whole area – putting millions of people at risk as up to 96 per cent of the country becomes suitable for transmission of the disease.

The picture is complicated however, because climate is only one factor affecting malaria transmission. In response to apocalyptic scenarios of malaria marching north into Europe and the US, many sceptics rightly pointed out that as recently as the nineteenth century malaria was common in England, the Netherlands, and the eastern United States. What eradicated the epidemics was not a change in the climate, but a change in material circumstances: economic growth and better healthcare. Hence, future projections of malaria transmission depend crucially not just on rainfall and temperature but economic and population scenarios. This explains the rather counter-intuitive result obtained by one 2004 study which found that hundreds of millions more people would be at risk of contracting malaria in a future scenario with lower greenhouse gas emissions. This was because even though the low-emissions scenario had less warming, it also saw lower economic growth combined with a bigger increase in population.

This complexity means that estimates for global changes in population at risk from malaria vary from 150 million *less* than today to 400 million more, all according to the same study. Either way it's bad news for Africa, however. Africa is the only continent which sees more people exposed to malaria in every single scenario, from 21 to 67 million people in the best and worst cases. The disease, like AIDS, will fuel a vicious circle in sub-Saharan Africa, where healthy and vigorous people succumb to malarial fevers all too regularly, taking productive workers out of the labour market and holding back agriculture. Countries in consequence remain mired in poverty, with their health systems under intense pressure. Whilst other parts of the world may be able to hold off malarial advances with stringent efforts, hospitals in Africa will remain jammed with sweating, writhing patients. And each time the rains pour more water into the sodden ground, clouds of mosquitoes will descend, and the cycle will begin again.

Paradise lost

To Arthur Conan Doyle it was the Lost World. A tabletop mountain in the heart of the Amazonian rainforest, frozen in time from the days when dinosaurs roamed the Earth, full of terrifying marvels that awaited the only team of scientists who dared approach. Conan Doyle's novel was based not on a fictional place, but on a remote corner of Venezuela where stunning flat-topped mountains every bit as fabulous and inaccessible really do rise like mist-shrouded vertical ships afloat in a sea of trees. Called *tepuis* by local people, waterfalls pour from their unclimbable heights, whilst their sheer sandstone cliffs are streaked with green where a few plants gain rootholds in cracks and gullies. Many have never felt the tread of human feet.

And Conan Doyle was not wrong in imagining strange creatures living atop them: from ferocious fer de lance snakes and toads that don't jump to jaguars and climbing rats, these isolated, precipitous summits are indeed unique. Vegetation on the flat tops of the mountains is luxuriant, and varies from rolling meadows to dense thickets of succulent bromeliads. On some tepuis 60 per cent of plants are found nowhere else on Earth. Most of the mountains are classed by ecologists as pristine, isolated as they are from human impacts like fire and deforestation, which threaten biodiversity elsewhere.

But this very isolation has bred vulnerability. The shape of the mountains, as a scattered archipelago of islands a thousand metres above a wider plain, ensures that most species can't migrate between them. Their flat tops mean that plants will be unable to move further uphill if climate change brings temperatures higher than they can tolerate.

Unfortunately, temperature rises above the plants' tolerance levels are exactly what lie in store as the world warms. According to a study by two Spanish biologists, more than a third of tepui endemic plants are likely to be wiped out. This would constitute a loss of biodiversity of worldwide importance, and the two scientists suggest that further urgent research be carried out to assess the possibility of saving as many species as possible through botanical gardening, seed or even DNA storing, with a view to reintroducing them into the wild once the climate has stabilised.

But given that many tepuis have never even been climbed, let alone explored and inventoried, the chances of a successful programme of this kind are remote. Moreover, a few specimens in a refrigerated laboratory would barely count as a species being 'saved'. For most, however, the death knell is already ringing as temperatures higher than for thousands of years begin to scorch the mountain tops. As the century progresses, the climate will move beyond anything these varied and unique plants and animals have experienced for millions of years, and many will not survive. Conan Doyle's world will truly be lost, this time for ever.

The fate of the tepuis illustrates that even the most isolated location cannot escape a worldwide change such as rising temperatures. From the deepest ocean to the frigid wastes of the Antarctic ice cap, climate change will be making an impact, imperceptible at first, but gradually becoming more and more disruptive as climatic zones shift and natural systems begin to tear apart.

On tropical coastlines, coral reefs will be bleaching annually by the time the world nears three degrees, with many whole reef systems already dead, and only their cooler outer fringes hanging on to a semblance of their original diversity. More than half of

Europe's plants will be on the Red List or already heading towards extinction, with those in mountain areas particularly vulnerable. In the Rockies and Great Plains of America, birds will face either drastic reductions in their habitat or moving 400 kilometres north to new areas. In north-eastern China, one of the last refuges of the Siberian tiger will be under threat as boreal conifer forests give way to hardwood invaders from the south.

The reason for this wave of destruction is simple: the different climates to which these species have become adapted through hundreds of thousands of years of evolution are going to disappear. A fascinating – if depressing – study, identifying exactly which areas are going to be the worst hit by 'disappearing climates', was published in April 2007, authored by a team led by John Williams, an assistant professor at the University of Wisconsin's Geography Department. It lists the Colombian and Peruvian Andes, Central America, the African Rift Mountains, the Zambian and Angolan Highlands, the Cape Province of South Africa, south-east Australia, portions of the Himalayas, the Indonesian and Philippine island archipelagos, and regions around the Arctic. In total somewhere between 10 and 50 per cent of the globe's surface will see its usual climate vanish altogether. Animals and plants that are adapted to these doomed climates will have nowhere to go: no place, anywhere on the Earth's surface, will still provide a suitable climatic habitat. Most depressing of all is the close association between those areas whose climates are projected to vanish altogether and global biodiversity hot spots. In other words, the places which experience the worst wipeout will be exactly those where today life flourishes in its most glorious abundance and diversity.

Believe it or not, these figures still underestimate the scale of the problem. Many species will see their climates shift towards the poles, requiring them to migrate rapidly in order to remain within the same climatic envelope – in other words, their climates will not disappear altogether, but will instead reappear hundreds of kilometres away to the north or south. Yet species' dispersal rates will be far outpaced by the speed of this change. For example, at the end of the last ice age, as the world warmed, trees and other plants were able at most to shift their ranges by 200 kilometres a century – and most moved much slower. Trace the distance between where a particular climate exists today and where it will exist in the future, and the gravity of the situation becomes apparent. Even if you assume a very generous 500 kilometres dispersal ability, animals and plants inhabiting between 40 and 85 per cent of the globe will see their climate disappear. Coincidentally, this is the same proportion of the globe's surface which the study calculates will experience 'novel' climates – those without an analogue for at least a million years.

Ecologists have a term for species whose habitat has already largely disappeared and whose populations have fallen so low that they are doomed to eventual extinction: the 'living dead'. Like Noah's Ark in reverse, dwindling groups of animals and plants, from forest-dwelling frogs to the polar bear, will be preparing to make a permanent exit from the world stage. We can get an idea of their numbers if we return to Chris Thomas and his colleagues' *Nature* paper from 2004, which suggests that between a third and a half of all species alive today will have joined the 'living dead' category by 2050 if global warming is over two degrees by that date.

It scarcely seems believable that life – in all its beauty, flamboyance and million-year resilience – could be under such a sudden and emotionless death sentence, that the world could never again witness the mating display of the bird of paradise or hear the haunting songs of the humpbacked whale. But the hard figures are there, compiled by experts working to rigorous scientific standards. Let no one doubt the consequences. The sixth mass extinction of life is well under way as global temperatures climb towards three degrees.

The Age of Loneliness has begun.

Growing food in the greenhouse

All plants have a thermal tolerance threshold, and the world's major food crops are no exception. Grains are particularly vulnerable to heat during flowering and setting seed, and temperatures over 30°C cause an escalating pattern of damage. According to John Sheehy of the International Rice Research Institute in Manila: 'In rice, wheat and maize, grain yields are likely to decline by 10 per cent for every 1 degree C increase over 30 degrees.' Over 40°C, yields are reduced to zero. With many areas in the tropics already close to or at this 30°C threshold, tropical yields in the three-degree world will be on a long downhill slide. Indeed, the global pattern will see a generalised shift in crop production away from the tropics and towards the more temperate higher-latitude regions, where cooler, wetter climes still prevail. There may still be enough food in these more northerly areas, but this tropical temperature crunch spells disaster for hundreds of millions of people.

As always, drought will play a key role. Agriculture in Africa's semi-arid tropics is largely rain-fed rather than based on irrigation, so is highly vulnerable to climatic shifts. North Africa could lose up to 20 per cent of its rainfall, whereas in southern areas decreases of 5 to 15 per cent will come right in the middle of the growing season. Agricultural modelling studies for the tropics as a whole project crippling declines in wheat, corn and rice production.

While crops in tropical areas will suffer because they are already close to their thermal tolerance threshold, those in higher latitudes may initially benefit because of longer growing seasons. However, once the global threshold of 2.5°C is crossed, even breadbaskets in the temperate mid-latitudes will begin to suffer as soaring summer temperatures leave crops suffering from a lack of water. As a major study for the 2007 IPCC report puts it: 'All major planetary granaries are likely to require adaptive measures by +2–3°C of warming, no matter what happens to precipitation.'

In the United States, southern areas closest to the subtropics will be hardest hit. Wheat and corn will suffer most losses as water supplies dry up. Droughts will be interspersed with heavier rainstorms, meaning crop losses can also be expected thanks to floods. Additional flood damages might total $3 billion in the US corn belt, according to one study. Pests and diseases also tend to benefit from warmer climes, necessitating a greater amount of pesticide use. Farmers will have to make big changes in the types of crops they grow and the amount of irrigation they use – difficult moves in western areas which will be pumping underground aquifers dry at the same time as run-off from snowmelt declines. Even in Canada increases in cereal crop production in the Prairies will be limited by water availability. Where water is plentiful, however, Canadian corn and soybeans could see big jumps in yields. Potatoes and winter wheat would also benefit. According to a study looking at the US and Canada: 'The range over which major crops are planted could eventually shift hundreds of kilometres to the north.'

However, with billions of people suffering drought and famine in the tropics and subtropics, the world food supply situation will become increasingly precarious, even with gains closer to the poles. The IPCC study sees net global food deficit beginning to drive up market prices once the 2.5°C threshold is crossed. Where the losses are worst in developing countries, widespread starvation becomes a real possibility.

With structural famine gripping much of the subtropics, hundreds of millions of people will have only one choice left other than death for themselves and their families: they will have to pack up their belongings and leave. The resulting population transfers could dwarf those that have historically taken place due to wars or crop failures. Never before has the human population had to leave an entire latitudinal belt across the whole width of the globe.

Conflicts will inevitably erupt as these numerous climate refugees spill into already densely populated areas. For example, millions could be forced to leave their lands in drought-struck Central American countries and trek north to Mexico and the United States. Tens of millions more will flee north from Africa towards Europe, where a warm welcome is unlikely to await them – new fascist parties may make sweeping electoral gains by promising to keep the starving African hordes out. Undaunted, many of these new climate refugees will make the journey on foot, carrying what they can, with children and old people trailing behind. Many of them will die by the wayside. Uprooted, stateless, and without hope, these will be the first generation of a new type of people: climate nomads, constantly moving in search of food, their varied cultures forgotten, ancestral ties to ancient lands cut for ever.

But these people may not be content to remain passive victims, for they will surely know that the world they inherit is not one that they have created. The resentment felt by Muslims towards Westerners will be tame by comparison. As social collapse accelerates, new political philosophies may emerge, philosophies which seek to lay blame where it truly belongs – on the rich countries which lit the fire that has now begun to consume the world.

4°

4

FOUR DEGREES

Death on the Nile

To say that Alexandria has a long history is something of an understatement. Named after its Greek warrior founder Alexander the Great, this Egyptian city soon grew to be one of the largest and most influential in the ancient world, second only to Rome in power and architectural glory. Its most famous building was the legendary 'Pharos of Alexandria' lighthouse, built in the third century BC and traditionally considered one of the Seven Wonders of the Ancient World, along with the Great Pyramid of Giza and the Hanging Gardens of Babylon. Hewn from snowy-white limestone and topped with solar mirrors for daytime lighting and a bright furnace at night, the Lighthouse of Alexandria was the tallest building in the world at the time, visible to sailors fifty kilometres out to sea.

Atop the tower rose a statue of Poseidon, Lord of the Seas, a god who was said to ride through the oceans in a chariot made from a giant clamshell and pulled by sea horses. Alexandrians worshipped Poseidon for understandably self-interested reasons: his power to cause storms and shipwrecks could send hundreds of men to a watery grave in seconds. The sea was the great power in the land.

Little remains of those times today – and Poseidon knows why. Unlike Rome, which managed to preserve much of its antiquity even as a thriving modern city grew above the ruins, only some Roman baths, a few catacombs and a single granite pillar remain of ancient Alexandria. Most of the rest was reclaimed by the sea as the Nile Delta, on which the ancient city was built, slowly subsided. Some of its finest treasures have been recovered from the seabed offshore after being submerged for centuries in Poseidon's watery grasp.

The waves are still encroaching, as global warming adds a further toll to the subsiding land. City authorities have to truck in huge amounts of sand from the inland desert near Cairo to keep the beaches from washing away, whilst massive breakwater fortifications have been constructed to keep the rising waters out of the city.

There is a lot at stake. Now a sprawling metropolis of 4 million people, modern Alexandria's two harbours form the busiest seaport in the country and support 40 per cent of Egypt's industry. Summer tourism adds another million visitors, making this the second-largest city in Egypt after Cairo.

But in the four-degree world, with global sea levels half a metre or more above current levels, Alexandria's long lifespan will be drawing to a close. Even in today's climate, a substantial part of the city lies below sea level, and by the latter part of this century a terminal inundation will have begun. A study conducted by scientists at the city's university suggested that by 2050 a rise in sea levels of 50 centimetres would displace 1.5 million people and cause $35 billion of damage. Indeed, as the sea begins to encroach across ever-wider parts of the Nile Delta, millions more will be evicted from their homes in other cities like Rosetta and Port Said. Beaches, wetlands and agricultural areas will all be submerged, devastating an area that is the heart of Egypt's economy. A part of the world that has always attracted conquerors, from Alexander to Napoleon, will be facing its final – and unvanquishable – foe.

Of course, Egypt will not be alone in suffering the effects of a global phenomenon like sea level rise. Further to the east, Bangladesh will be losing a third of its land area, displacing tens of millions from the fertile Meghna delta. In Boston, USA, storm-surge flooding from higher sea levels could inundate even the city's central business district by 2075, causing estimated damages of $94 billion. Down the coast in New Jersey a 60-centimetre rise in sea level would flood 170 square kilometres of land, whilst a rise of double that would drown more than 3 per cent of the state, including some of the most densely populated coastal areas.

Low-lying and deltaic cities from Mumbai to Shanghai are just as threatened as Alexandria and Boston. New York, London and Venice will only be saved if huge amounts of money are ploughed into new and ever-higher defences against floods, as we saw earlier. Like today's New Orleans, coastal cities of the future may gradually become fortified islands, largely below sea level and under siege from all sides by the advancing waters. Such a strategy would protect trillions of dollars' worth of real estate, but it would also bring dangers: as New Orleans fatefully experienced, one serious storm can bring down a vulnerable city in a matter of hours, putting many thousands of lives at risk. Rebuilding a city may be an option after the water is pumped out, as long as insurers are willing and able to cough up the necessary sums. But who will pay to rebuild a city twice? Or three times? In the very long term, the only solution will be for hundreds of millions of coastal dwellers to retreat inland, as civilisation's map is continually redrawn with constantly changing geographical boundaries.

The pressure on societies will be immense. Inland cities will face a constant stream of refugees from coastal areas, with thousands – and perhaps millions – arriving all at once when major storms hit. (Again, this is not conjecture: as I write hundreds of thousands remain displaced by Hurricane Katrina, across Texas and other southern states, two years after that disaster.) As economic shocks due to direct losses, social instability, declining public confidence and insurance payouts cascade through the financial system, the funds to support displaced people and build new living areas will become increasingly scarce. Given that sea level rise is an irreversible process, which will take millennia to stabilise even if greenhouse gas levels are controlled, refugee cities may themselves become endangered and submerged in decades and centuries to come. In today's cities, many of the finest and most treasured buildings are hundreds of years old; in the future coastal buildings may last only a matter of decades before the rising waters begin to approach once again.

Island nations face the greatest threat: whilst coral atolls will disappear completely, mountainous islands like Fiji and Barbados will suffer as their coastlines shrink and refugees struggle to make a living on denuded hillsides and higher cliffs. In many islands the most fertile lands often lie just above sea level, so food and water supplies are hit as salt-water penetration poisons aquifers and kills crops.

Uncertainties here are great: if the Antarctic ice sheets remain stable, much can probably be salvaged in a slow and measured retreat. If, on the other hand, the great ice sheets continue to respond rapidly to climate change – as they are already doing now – many metres of rapid sea level rise are on the cards. As I showed in the previous chapter, judging from palaeoclimatic evidence from the Pliocene, eventual rises of 25 metres from Greenland and Antarctica are pretty much inevitable once global temperatures pass two degrees. Even spread out over many centuries, this would stretch the human capacity for adaptation. Coastal zones will be in a constant state of flux, their human inhabitants insecure and threatened as the waves roll closer each year.

Not for nothing did the ancient inhabitants of Alexandria put serious effort into placating Poseidon, the powerful deity who could summon up storms and sink ships with a single wave of his trident. Coastal residents of today have lost their respect for the sea, and expect governments and city authorities to be able to protect buildings which are built much too close to the shore. Most of us assume that the ocean can be tamed and controlled, just as big dams have tamed once great rivers like the Yangtze and Colorado. But Poseidon is angered by arrogant affronts from mere mortals like us. We have woken him from a thousand-year slumber, and this time his wrath will know no bounds. Modern Alexandrians will be among the first to flee before his unstoppable oceanic advance. This conqueror may be fought, resisted, even temporarily checked, but he can never be defeated.

The heart of Antarctica

On the far side of the world from Egypt, sprawled across the South Pole, lies Poseidon's biggest armoury – Antarctica. So far, the great Antarctic ice sheets have been slow to respond to humanity's interference with the global climate, isolated as they are from the rest of the world by powerful cooling winds and currents. But now, to paraphrase the former British Antarctic Survey director Chris Rapley, the slumbering giant is awakening. The West Antarctic Ice Sheet has one great weakness, an Achilles heel that was recognised as long ago as 1978 by prescient glaciologists. Unlike Greenland's ice cap, which is firmly anchored on a continental landmass, much of the base of the West Antarctic Ice Sheet is grounded below sea level – and that makes it vulnerable to collapse.

The first signs of change are all around. The monumental

glaciers which drain ice from the centre of the continent have begun to speed up and retreat, and the great ice sheet is losing somewhere between 90 and 150 cubic kilometres of ice every year. This thinning has also propagated far inland, lowering the ice sheet by a metre or more as far as 300 kilometres away from the coast. Scientists have subtracted this ice loss from the annual snowfall total over west Antarctica and discovered that the glaciers are adding 0.14 millimetres to global sea levels each year – a small but growing new weapon in Poseidon's arsenal.

As the warm oceans erode west Antarctica's edges, the stability of the whole ice sheet is being called into question. Water flows downhill, and ice floats on water. Given that much of the centre of the great ice mass is even deeper below sea level than are its edges, penetrating seawater could in theory lift much of the ice cap off its seabed foundations, adding 5 metres to global sea levels in just a few decades. The process would be rapid not just because of these physical dynamics, but because water is a very efficient conductor of heat: warm water melts ice a lot faster than warm air, and the warming Southern Ocean is a dagger pointed at the heart of Antarctica.

The West Antarctic does have some last lines of defence. Two gigantic ice shelves – the Ross and Ronne, both the size of Texas – stand like armoured fortresses against any invasion of seawater under the main ice sheet. Both, although floating, have sheer ice ramparts between 200 and 400 metres thick at their northern edges where they meet the swells of the open ocean. Both are safe for now because they are far outside the surface melt zone: their temperatures stay well below freezing all year round.

Or so everyone thought, until recently. In May 2007 NASA scientists reported that the agency's QuikSCAT satellite had found clear evidence of the first widespread melting ever detected in Antarctica. In January 2005, at the height of the austral summer, an area of ice and snow as big as California had begun to thaw. The melting wasn't just concentrated around the coast – it penetrated 900 kilometres inland, crept 2,000 metres up Antarctic mountainsides, and even came within 500 kilometres of the South Pole itself. Temperatures rose as high as 5°C, and remained above freezing for a week. This sudden Antarctic melt may be a one-off – up until March 2007 no further thawing had been detected. But, says Son Nghiem, a scientist involved in the study, 'It is vital we continue monitoring this region to determine if a long-term trend may be developing.'

A large proportion of this melt took place on the northern edge of the Ronne Ice Shelf – suggesting that one of west Antarctica's most important lines of defence may soon begin to crumble. Evidence of its likely eventual fate has emerged from further north in the continent, along the Antarctic Peninsula, which juts in a spine of ice-clad mountains out towards the Patagonian tip of South America. Glaciologists have long argued that a loss of floating coastal ice shelves could result in faster glacier flows upstream, just as a cork popped out of a bottle allows the champagne to flow freely. Three major ice shelves on both sides of the Peninsula have now disintegrated: the Wordie, Larsen A and Larsen B, the latter collapsing spectacularly in a matter of days in March 2002. In each case, the glaciers which act as tributaries into the ice shelves have accelerated their flow of ice into the sea, just as predicted.

A global warming of four degrees would be more than enough to allow the melt line to creep across both the Ross and the Ronne ice shelves, fatally damaging their integrity as meltwater prises open gaps in the ice. If either breaks up, just like the Larsen and Wordie ice shelves already have further north, then nothing will stand in the way of total collapse for the entire West Antarctic sheet and rapid inundation for the world's coastlines.

But west Antarctica's demise would be just the first battle in a much longer war. Its much larger neighbour, the East Antarctic Ice Sheet, is 4 kilometres thick in parts, and holds enough water to raise global sea levels by more than 50 metres. That, together with Greenland's collapse, would put me on the coast at my home in Oxford, located on the 65-metre contour line. The UK would be split into an archipelago of hilltop islands, and the world's continental coastlines would be barely recognisable from today.

The East Antarctic has an even more formidable line of defence than the West – a line of high mountains, the Transantarctic range. Seawater could of course never cross this mountain range, however hot it gets. But the East Antarctic sheet might be vulnerable via the back door, where it too is anchored below sea level. Most scientists don't even realise it, but these little-known submarine beds extend to the very centre of the ice sheet. I am not suggesting that collapse would happen instantaneously – in fact it would take centuries, and probably millennia, to melt all of the Antarctic's ice. But the destabilisation of both major Antarctic ice sheets could yield sea level rises of a metre or so every twenty years – far outside the adaptation capacity of humanity.

So will it ever happen? Again, geological evidence could be key. When the Earth was last four degrees warmer, there was no ice at either pole. Global warming of this magnitude would eventually leave the whole planet without ice for the first time in nearly 40 million years.

Capitalism with Chinese characteristics

Climate change is not the only major ecological challenge facing humanity, although it is undoubtedly the most serious and urgent. Global warming is joined by other mounting threats – including population growth, soil loss, fossil aquifer depletion, and the wholesale destruction of ecosystems – each of which has the potential to escalate into a major survival crisis for modern civilisation. Nowhere is this more apparent than in China, which is industrialising at breakneck speed, transforming itself from a largely peasant nation into an economic powerhouse in less than two decades. The country's leaders and population have been heading towards a Chinese-variant hypercapitalism ever since Chairman Mao breathed his last, and the economic reformers led by Deng Xiaoping came to power and quickly declared that 'to get rich is glorious'.

Glorious it may be for the new millionaires who cruise through the glittering canyons of Shanghai and Beijing, flaunting their new prosperity with celebrity-style conspicuous consumption. Glorious it may be too for the tens of millions of ordinary Chinese who no longer live in dire poverty, and own substantial capital for the first time. But for China's ecological capital, economic growth has been utterly disastrous. A fifth of the country's native biodiversity is now endangered. Three-quarters of its lakes are polluted by agricultural or industrial run-off, whilst the Yellow River is depleted and virtually toxic along much of its lower reaches. Almost all China's coastal waters are polluted by sewage, farm pesticides and oil spills, causing on average 90 poisonous red tides per year. Approximately 15,000 square kilometres of grasslands are annually degraded by overgrazing and drought. Acid rain falls on a quarter of its cities. Three out of four urban residents breathe air which falls below minimum health standards. In Hong Kong's 2006 marathon, for example, several runners were hospitalised and one died after completing the course in persistent smog.

Because of its sheer size and population, China is on a collision course with the planet. The country's oil use has doubled in the last ten years, and if the Chinese by 2030 use oil at the same rate as Americans do now, China will need 100 million barrels of oil a day. However, current world production is only around 80 million barrels per day, and is unlikely to rise much further before the 'peak oil' point is reached. There simply isn't enough oil in the ground to bring Chinese consumption up to Western levels – the global resource buffer is already being hit.

Similarly for food: as the Chinese diet becomes increasingly rich in meat and dairy products, more grain is needed. By 2030, if Chinese consumers are to become as voracious as Americans, they will use the equivalent of two-thirds of today's entire global harvest. If Chinese car ownership were to reach current US levels of three cars for every four people, China's automobile fleet would number over 1 billion by 2030 – substantially more than the entire current world fleet of 800 million.

In almost every sector of resource use, China's ascension to Western consumption standards will clearly demand far more than the Earth can provide. Indeed, if every Chinese were to live like an American, it would double the human environmental impact on the planet, an impact which has already moved far beyond sustainable levels. Even forgetting about climate change, China's get-rich-quick dream would quickly become a global nightmare.

But climate change cannot be forgotten, and its impacts on China will be severe, worsening the shocks as the country and our planet hit the ecological buffers. One study conducted by the UK and Chinese governments suggests that by the latter third of the twenty-first century, if global temperatures are more than three degrees higher than now, China's agricultural production will crash. Yields of staple crops like rice, wheat and maize will decline by nearly 40 per cent; perhaps more if water supplies for irrigation run out. China will face the unenviable task of feeding 1.5 billion much richer people – 300 million more than now – on two-thirds of current supplies.

Of course world markets could in theory supplement the gap, but agricultural breadbaskets will be suffering declines right around the world by this time, with whole areas knocked one by one out of production. These include substantial parts of western North America, the Pacific coast of South America, southern Africa and the western half of the Indian subcontinent, all because of the declining river waters and spreading deserts detailed in earlier chapters.

In addition, none of the continent of Australia – except perhaps the extreme north and Tasmania – will be able to support significant crop production in the four-degree world because of heatwaves and declining rainfall. In India, precipitation is projected to increase in most areas because of a more intense summer monsoon, but with land temperatures soaring to 5°C or more above current levels, it will simply be too hot for most crops to survive. Moreover, faster evaporation in the hotter climate may actually make soils drier in many areas. In western areas of the subcontinent, particularly in Rajasthan, Punjab and neighbouring Pakistan, already arid areas get drier still, compounding the water emergency arising from the deglaciation of the Himalaya and Karakoram mountain chains. All of these regions will be haemorrhaging people in the biggest human migration ever seen, with hundreds of millions on the move in search of food and water.

A global analysis conducted for the 2007 IPCC report identifies several 'hotspots for future drought': south-western North America, Central America, the Mediterranean, South Africa and Australia. During the winter months, south-east Asia is a drought hot spot, whilst the Amazon, Siberia and part of West Africa suffer most during the summer. Even areas that get more average rainfall, such as the higher mid-latitudes, will see this extra wetness coming in winter, outside the main growing season when crops cannot benefit as much. The study sums it all up with an ominous phrase: 'worldwide agricultural drought'.

It is plausible that new areas of production in sub-polar regions of Canada and Russia will be able to cover some of the shortage, though warmer temperatures across thawed tundra do not equate to higher rainfall or decent soils. It is also likely that new techno-logical developments, with more drought-resistant crop strains, could help to stave off disaster for a while, as could the fertilisation effect of higher CO_2 levels in the air. But none of this can make up for the loss of most of the planet's key agricultural areas, and it is difficult to avoid the conclusion that mass starvation will be a permanent danger for much of the human race in the four-degree world – and possibly, as suggested previously, much earlier. With major global breadbaskets dusty and abandoned, rising demand will be chasing rapidly diminishing supply.

How this food crisis might play out in different areas is impos-sible to predict. History, however, is littered with the ruins of soci-eties that collapsed once their environments became overstretched and their food supplies endangered. The case of the Mayans is one of the better known, but in China too, early civilisations rose and fell in response to fluctuations in rainfall and drought. The ancient Harappan civilisation of the Indus River valley was also likely extinguished by a particularly severe drought 4,200 years ago.

A similar cause seems to have precipitated the disappearance of the various Middle Eastern kingdoms which in their time must have imagined themselves unassailable, just as we do today. One is reminded of Shelley's poem 'Ozymandias', about the shat-tered statue of a long-dead king who sneers in carved words at passers-by: 'Look on my works, ye Mighty, and despair!' The poem ends:

Nothing beside remains. Round the decay
Of that colossal wreck, boundless and bare
The lone and level sands stretch far away.

The sands of Europe

All of the civilisational collapses mentioned above took place as a result of comparatively small changes in climate, changes which will be dwarfed by the massive shifts we can expect to see in the century ahead. If just a few tenths of a degree did for the Maya and the Harappans, imagine what ten times that might do for our fragile and interconnected world today. In some ways, the situation is even worse now because this time our ecological crisis is truly global: when the Mayans had deforested their local area and exhausted their food supplies, the ragged survivors of the resulting wars and chaos at least had somewhere else to flee. Migration is the traditional human adaptation to crisis, but this time there will be nowhere to hide. Civilisational collapse, like the blast wave of a neutron bomb, will sweep around the globe.

Even worse, this calamity is overtaking a planet whose natural defence mechanisms have already been severely damaged by human activity. In their natural state, ecosystems play a vital role in regulating the climate, thereby keeping the Earth habitable for life. Plankton, for example, release gases which help to promote cloud formation, whilst Amazonian rainforest trees generate their own thunderstorms by recycling water across vast distances. In the long term, life in the oceans helps keep atmospheric carbon dioxide at tolerable levels by sequestering carbon in ocean sediments, which then form carbonate rocks like chalk and limestone. On land, vegetation accelerates the chemical weathering process of soils, which also locks up carbon.

But these natural ecosystems have been drastically reduced in extent. The majority of the planet's fertile soil has been stripped of trees and grasses and appropriated for farmland to feed humans. The oceans have been sucked dry of everything from cod to krill by gigantic factory ships. In total, humans now appropriate over 40 per cent of the entire photosynthetic productivity of the planet, leaving the rest of nature to scrape along at the margins, in the places which are too hot, too cold, too high or too deep to yet be useful to us.

In his book *The Revenge of Gaia*, James Lovelock speaks of this as a 'double whammy': like the engineers at Chernobyl who inadvisably turned up the heat after disabling their reactor's safety systems, we have disabled the Earth's thermoregulation systems by cutting down forests and polluting oceans – just at the time they are most needed. The human experiment on the planet's climate – of turning up the heat with billions of tonnes of green-house gases, whilst at the same time wiping out most natural ecosystems that help to regulate the climate – will have the same predictable effect on the Earth as the Soviet technicians had on Chernobyl's reactor core when they tried the same experiment, Lovelock suggests. It will lead to meltdown.

By the time global temperatures are climbing towards four degrees, this meltdown process will be well under way. Greenland's ice sheet will be shrinking year on year into the centre of the landmass, spilling vast amounts of water into the rising seas. Dramatic changes, as described earlier in this chapter, will also be under way in Antarctica. The Atlantic circulation – if it survives current blips – will finally slow down and stop. (This would come too late to make Europe colder than it is now; it might just moderate some of the more extreme warming in places like the UK.) The world's weather will go increasingly haywire, with wilder storms mobilising undreamt-of ferocity as they strike ever-larger areas. The long months of summer will get longer still, as soaring temperatures reduce forests to tinder and cities to boiling morgues.

In southern Europe, new deserts will be spreading. Shelley's poem, which today evokes the Middle East, may summon to mind Italy, Spain, Turkey or Greece in the four-degree world. Scientific studies exhibit near-unanimity in projecting drier climates with far hotter temperatures for the Mediterranean fringe of Europe. One recent modelling exercise, conducted by scientists based in Sweden and Finland, projects rainfall declines across the entire area – with up to 70 per cent of summer rainfall disappearing. A second study predicts that heatwaves could be up to 65 days longer across all the major tourist areas of the Mediterranean: Spain, Portugal, southern France, Italy, Greece and Turkey, with ripple effects reaching as far as southern Russia and Ukraine. Another, published in June 2007, projects a 200 to 500 per cent increase in the number of dangerously hot days, with France and Spain the worst-affected areas. Worryingly, the epicentre of the heat increase is exactly the zone most badly affected by the 2003 heatwave, which killed 15,000 people in France. Even so, the heatwaves projected for the future will be far hotter even than the extremes of 2003. In essence, hot subtropical climate zones now located in North Africa will have spread north into the heart of Europe.

Even in the more temperate climes on the far side of the Alps, the mercury will be hitting new heights. In Switzerland, as heatwaves scorch mountains and valleys once famous for their verdant lushness, the extreme summer of 2003 will look cool by comparison. July and August temperatures could hit 48°C, more reminiscent of Baghdad than Basel. Wildfires will sweep up Alpine slopes, and water supplies drop precipitously as the remaining mountain glaciers waste away on the highest peaks. Even in England, where 11 August 2003 set a new all-time high of 38°C, summer heatwaves could see temperatures in London and the home counties reach a searing 45°C – the sort of climate experienced today in Marrakech, Morocco. Summer droughts will put the densely populated south-east of England on the global list of water-stressed areas, with farmers competing against cities for dwindling supplies from rivers and reservoirs.

Russia's harsh cold will be a distant memory as December-to-February temperatures rise on average by 7 degrees, leaving most winters without snow even in formerly chilly eastern Europe. Snowfall totals are expected to plummet by 80 per cent or more across the continent, with only the interior part of Scandinavia's far north still receiving reliable winter snows. The new snowless regime will exacerbate winter flooding, as more precipitation comes in the form of rain and runs straight into rivers. Without the snow melting slowly and releasing water late into spring, summers will become drier, leading to temperatures on the continent rising by up to 9°C above present. This nine degrees figure is just the *average* – in extreme years heatwaves will be almost unimaginably severe. The Caspian Sea, which gets most of its run-off from Russia, is projected to drop in level by nearly 10 metres, lower than it has been for at least 25 centuries. With these dramatic weather shifts, it is scarcely surprising that north-eastern Europe and the Mediterranean have been identified as the two most prominent future climate change 'hot spots'.

Summer, not winter, will be the season Europeans dread in the four-degree world. As in any southern US city today, airconditioning will be mandatory for anyone wanting to stay cool. This in turn will put ever more stress on energy systems, which could pour more greenhouse gases into the air if coal- and gas-fired power stations ramp up their output, hydroelectric sources dwindle and renewables fail to take up the slack. Still more pollution may come from seawater desalination plants in desertifying countries like Portugal and Spain, unless these are powered exclusively by solar energy. But controlling further greenhouse gas emissions may be the last thing on any politician's agenda, desperate as they will be for fresh water to keep cities habitable and prevent agricultural collapse.

The summer-long heatwave of 2003 and its associated mass mortality gives some idea of how a warmer Europe might look, but with every summer as hot or hotter than 2003 it is difficult to see how societies would cope in the longer term. Instead of the Mediterranean being the densely populated place it is today, whole areas may be abandoned – their residents flocking north, to overcrowded refuges in the Baltic, Scandinavia and the British Isles. Whether the continent can hold together will depend on the strength of its institutions and the determination of its peoples to survive the emergency. But with habitable areas becoming more and more crowded, conflict and chaos may come sooner rather than later even in temperate, civilised Europe.

Into thin air

Above the tree line, only a few stunted shrubs break up the monotony. Clumps of grass cluster on the barren lower slopes, but above them nothing interrupts the moonscape of rock and scree. Goats pick their way between the few traces of green, munching on anything they can find: moss, bark or thornbush; nothing is beyond them. Ridge after ridge soars away into the distance above, browns, greys and reds merging into the horizon. Only the faintest smears of snow cling to the highest summits, whilst in the corries and high valleys below, curved ridges of moraine rubble mark the former extent of long-forgotten glaciers. The lower valleys are cut by chasms where flash floods have torn debris from the mountainsides, leaving huge boulders isolated amidst wide fans of alluvium. But the chasms are dry now – only an occasional muddy pool remains from the most recent rainstorm. And all around, in the shimmering heat haze, is bare rock. The mountainsides are naked.

This scene might be familiar to anyone who has trekked in the Moroccan High Atlas, the 4,000-metre-high mountain chain that divides the Mediterranean coastal plain from the wastes of the Sahara. It might also be familiar to anyone trekking in the Alps later in the twenty-first century, when what is today a North African climate becomes firmly established in Europe. By the time global temperatures approach four degrees above today's levels, snow will be a rarity at Alpine elevations below 1,000 metres – places which can currently expect between 50 and 100 days of snow cover each winter. At 2,000 metres two months will be wiped off the snow season, with the amount of snow accumulating during the winter cut by half. Even as high as 3,000 metres – where snow often lies all the year round in today's climate – a third of it is expected to melt away by winters in the 2070s. Even more striking, glaciers will vanish from the majority of even the highest peaks, making the Alps almost completely devoid of ice for the first time in millions of years. Only tiny remnant glacier patches may hang on at the top of the highest 4,000-metre peaks like Monte Rosa and Mont Blanc.

Heatwaves will even hit in the winter months, pushing temperatures up to 20°C between December and February, and melting snow right up in the highest peaks. Although no one will suffer heat exhaustion in twenty-degree winter temperatures, the effects of these winter heatwaves on landscape and society can be dramatic. Sudden rises in temperature make upper snow slopes unstable, raising the risk of avalanches dumping millions of tonnes of suffocating wet snow on villages down in the valleys. Dramatic melts send torrents of muddy water down the hillsides, causing deadly flash floods and washing out buildings and bridges with unstoppable mudflows. Plants are tricked by the temperatures into beginning early spring growth, only for buds and tender leaves to be killed off by renewed cold in days and weeks to come.

Given their crucial role as the 'water tower of Europe', the impacts of snow cover changes and glacial disappearance in the Alps will propagate throughout the continent because of their effects on major rivers like the Rhine and the Danube, which have their headwaters in the mountains. In Switzerland itself, 60 per cent of current electricity comes from hydroelectric power generation, an energy source which may fail during the summer months as streams and rivers run dry. As with the American and Canadian Rockies, the problem is one of timing: even if the total amount of yearly precipitation remains similar to now (and there's no guarantee of that), more of it falling as rain than snow in the winter means that peak stream flows occur earlier in the year, reducing the amount that is available during the summer for human use.

With neither snowmelt nor rain, vegetation will wither, turning the green landscape into baked-earth browns as the grip of drought intensifies. As in the High Atlas of today, no glaciers will grace the Alps of the future, and no greenery will break the monotony of rock on the highest slopes. The mountains themselves will be neither higher nor lower than today, but their character will have changed utterly. Mountain residents – plant, animal and human – too will have to change if they are to survive. Most important of all, as Europe swelters in summer heat, the continent's water tower will run dry.

Blighty gets a battering

As people migrate north from the searing heat of Saharan southern Europe, Britain's relative cool will be making these crowded islands one of the most desirable pieces of real estate on the planet. Even if drought is a problem in the south and east, the north and west seem likely to continue to receive reliable watering thanks to Atlantic weather systems. That would be fine, were it not for the increasingly ferocious nature of these Atlantic storms. With cyclone tracks moving towards the poles as subtropical dry belts shift outwards from the equator, the UK will lie right at the centre of the action – as storms which no longer drift over the drier Mediterranean instead track over northern Europe. Scotland – which already sees the strongest winds and heaviest rainfall in Britain – will be particularly affected. With deeper lows and stronger winds, each depression will pack a more severe punch than even the fiercest winter gales in the current climate.

This means greater storm-surge flooding in areas exposed to westerly gales, more coastal erosion and damage to infrastructure from strong winds. The costs of this damage could increase by up to 37 per cent above today's levels in both the UK and Germany, bankrupting insurers with recurring losses in the tens of billions of euros. However – perhaps surprisingly – even though individual storms are expected to get stronger, there may actually be fewer of them. That will at least give emergency services a chance to move injured people and conduct urgent repairs to roofs, power lines and flood barriers before the next cyclone moves in.

Needless to say, these storms will also bring heavier rainfall. With more cloudbursts will come more extensive floods, but the hardest-hit part of the UK in this case will not be Scotland, but the highly populated south and east of England. Although some of this extra water will be vital to top up reservoirs sucked danger-ously low during hot summers, too much of it will come in heavy bursts, running quickly off the land and streets rather than soak-ing more gently into the soil to recharge groundwater supplies. Communities will struggle to cope as summer drought gives way to winter downpours, sending floods surging through towns and villages as rivers burst their banks. Even in the current climate, over 4 million people and 2 million properties are at risk of flooding in the UK; these risks are expected to quadruple in a world nearing four degrees of global warming, with annual flood damage costs nearing £30 billion every year.

As flood plains are more regularly inundated, a general retreat out of high-flood-risk areas is likely – a reverse of the current trend, where housing estates are constructed often without consideration of flood risk. Millions of people will as a result lose their lifetime investments in houses which become uninsurable and therefore unsaleable. 'Flood blight' may become common parlance in the housing business. According to the UK government's Office of Science and Technology, the Lancashire/Humber corridor is expected to be among the worst-affected regions, as are the Thames Valley, eastern Devon, and towns around the already flood-prone Severn estuary like Monmouth and Bristol.

Many of these are the same areas projected to suffer most from accelerated coastal erosion, due to a combination of rising sea levels and stormier weather. The entire English coast, from the Isle of Wight in the south to Middlesbrough in the north-east, is classified as at 'very high' or 'extreme' risk in this future climate, as is the whole coastline of Cardigan Bay in Wales.

Another change in winter weather will be noticed by everyone, although its direct impacts are less clear. Snowfall – something eagerly anticipated each winter by generations of schoolchildren – is set to disappear almost completely from the British yearly calendar. In the lowland south and east of England, lying snow will be virtually unknown in the four-degree world. Even in the Scottish mountains, snow cover will drop by more than half. It may still occasionally snow in English winters, but snowfall will be a freak event, like marble-sized hail or ball lightning. Kids may still be able to build a snowman if they want to – but they'll have to make a January trek up to the summit of Ben Nevis first.

A buried message in Texas

On the south-central Texas ranch of T.D. and Billie Hall, an unusual cave lies hidden in the undulating limestone landscape. Rocks litter the entrance, and a lone hackberry tree stands guard as if to ward off intruders. All around, the terrain is arid, with exposed bedrock in places, and just a few evergreen oaks dotted around to break up the rolling savannah. A few kilometres to the south, one fork of the Guadalupe River winds its way through the stony plateau towards the Gulf of Mexico. A few cattle wander about, but stocking rates are low – here the thin soils cannot support big grazing herds like the prairies do further north.

Inside Hall's Cave, Jennifer Cooke, a University of Texas PhD student, spent many weeks making detailed forensic excavations, before coming up with a surprising discovery. Buried in the sediments at the bottom of the cave were teeth and bones – not from some long-murdered cowboy, but from gophers and other burrowing mammals that had foraged in the ground above the cave over twenty thousand years earlier. The discovery was surprising because none of these animals live in the vicinity now: the soils are simply too thin. What Cooke began to piece together was a story of major climate-driven soil loss, long before humans occupied the landscape, which turned this once fertile plateau into the stony, arid wasteland it is today.

The erosion agent was a familiar one: rain. In response to pounding thunderstorms – interspersed with increased droughts – the soils which had built up over previous millennia simply began to wash away. The warming post-ice-age climate apparently triggered more intense cloudbursts, hammering the drying landscape and washing sediment off the land surface and down into the rivers. Cooke also found thick layers of the old soil in the cave itself, where they had been deposited by the heavy rain. Her PhD thesis was later shortened into a paper in the journal *Geology*, where she warned that future climate change – with 'greater summer aridity and more frequent, intense rainfall' – could again trigger the kind of events which destroyed fertility across a massive area of Texas many thousands of years ago.

All of this might sound rather ominous to anyone familiar with studies of likely future climate based on computer model outputs. One recent paper, focused on the United States and published in *Geophysical Research Letters* in late 2005, projects that more convective rainfall is likely, with a higher rainfall intensity across the whole country. (Convective rainfall tends to come in small, violent cloudbursts like thunderstorms, as opposed to the gentler rain associated with the passage of frontal depressions.) 'Heavy precipitation events should become more frequent,' the authors warn. Rainfall, moreover, 'should become more episodic and with larger daily amounts', though with longer dry periods in between.

The entire land surface of the globe is affected by these changes in rainfall in the four-degree world. In central and northern Europe heavy precipitation increases significantly, particularly over wintertime Scandinavia. These increases are accompanied by decreases in the south of the continent, causing drought and desertification, as we saw earlier. Winter cyclones get more intense and destructive in both hemispheres, though their frequency diminishes. Numbers of tropical hurricanes are not expected to change much, but their intensity – as we saw in the previous chapter – is projected to increase strongly. One modelling study of the eastern Australian region found that the number of strong storms increased by more than 50 per cent, with more intense cyclones hitting further south – even putting Sydney in the firing line (though it may be grateful for the rain). In Korea, rainfall rises by a quarter – but land temperature increases of 6°C also drive up evaporation, resulting in a drier land surface than before.

Indeed it is the sheer temperature rise which begins to dominate over everything else in the four-degree world. Heatwaves of undreamt-of ferocity will scorch the Earth's surface as the climate becomes hotter than anything humans have experienced before throughout their whole evolutionary history. As we saw earlier, temperatures in Europe will by this time resemble the Middle East more than our more usual temperate climes. The Sahara will have crossed the Strait of Gibraltar and be working its way north into the heart of Spain and Portugal. Even where cultivable soils remain, heavy cloudbursts accelerate erosion, converting once fertile fields into gullied badlands, just like the Texas plains. With world food supplies crashing, humanity's grip on its future will become ever more tentative.

Siberian roulette

In previous chapters we saw an Arctic which was physically unravelling. With global temperatures nearing three degrees, summer sea ice was reduced to a remnant patch at the pole and the far north of Greenland. Now, with temperatures past the three-degree mark and moving towards four degrees, even the more conservative computer models predict that the sea ice disappears completely, and – for the first time in at least 3 million years – the summer North Pole sees nothing but open ocean. Even during the long dark nights of the polar winter, much of the ice fails to re-form. Temperatures across the region soar to 14°C higher than current levels during the winter months.

In the continents that surround the Arctic Ocean, similarly drastic changes are under way. The southern boundary of permafrost shifts hundreds of kilometres to the north. Its area shrinks from over 10 million square kilometres today to a mere 1 million square kilometres by the end of the century. As the process accelerates, large areas of Siberia, Alaska, Canada and even southern Greenland fall into the melt zone, where unstable soils shift and collapse underneath roads, houses and other infrastructure. In the Russian Far East, towns like Yakutsk, Noril'sk and Vorkuta find themselves built on quicksand. The Trans-Siberian Railway suffers extensive subsidence along its track, with changes even endangering a nuclear power plant at Bilibino. All around the Arctic Ocean, increased erosion from storms and rising sea levels destroys shoreline villages and settlements.

Arctic ecosystems find themselves in a state of advanced dislocation. Unexpected fires and insect attacks wipe out forests far to the north of the Arctic Circle. With snow coverage declining by 20 per cent across the whole region, animals like voles and lemmings – which forage in tunnels under the snow during the winter – face decline. Even where snow does cover their feeding grounds, occasional thaws, rain on snow and ice crusts all reduce the insulating properties of the snow layer and kill the tiny animals with cold. Because lemmings are prey for arctic foxes, snowy owls, weasels, skuas and ermine, all these predators in turn face food shortages.

As the animals struggle to survive, all around them the landscape is changing. Lakes are draining and rivers altering course, with 30 per cent more fresh water draining into the Arctic Ocean from extra rainfall and thawing ground. As the snow and ice retreats, trees are growing in the mossy tundra. New bogs are appearing in ground which was once frozen as solid as concrete.

And out of these thawing Arctic soils a new threat has arisen, one of the most dangerous positive feedbacks of all, one which sees the effect of climate change in the Arctic ricochet around the rest of the planet with increasing force and destructiveness. Like the positive feedback from soils, this threat comes from the fact that global warming accelerates the release of greenhouse gases from the ground, thus further accelerating global warming in an ever-tighter spiral.

Around 500 billion tonnes of carbon are currently estimated to be locked up in permanently frozen Arctic soils. Once thawing begins, much of this carbon will begin to escape. In places where the draining of lakes and marshes allows soils to dry out, it can enter the atmosphere directly as carbon dioxide, as soil bacteria begin to break it down. Where the soils remain too wet for oxidising decomposition, anaerobic bacteria move in and produce vast quantities of methane – an even more dangerous greenhouse gas than CO_2 due to its more powerful short-term effect on the climate. In other areas carbon can dissolve directly into water, and be released as CO_2 from rivers, lakes, and the Arctic Ocean. Says Phil Camill, a US-based ecologist studying the melt rates of Canadian permafrost: 'We are unplugging the refrigerator in the far north. Everything that is preserved there is going to start to rot.'

This rotting effect is already being observed – though at a more limited extent – in the current climate, where permafrost degradation is already advancing across the Arctic. Californian geographers Karen Frey and Larry Smith spent three years between 1999 and 2001 trekking across remote parts of western Siberia to take nearly a hundred samples of 'dissolved organic carbon' from streams and rivers. Once the water samples were analysed in the laboratory, a clear pattern emerged: Frey and Smith discovered that watercourses draining defrosted peatland had far higher carbon counts than rivers running over still-frozen permafrost areas. When they went on to plot future warming scenarios onto maps of current permafrost distribution in Siberia, they found that major melting could produce a staggering 700 per cent increase in carbon release.

Hundreds of kilometres away in the Abisko region of subarctic Sweden, scientists were at the same time trying to measure how much extra methane is released by thawing bogs. Emissions from one recently defrosted mire were estimated at between 20 and 60 per cent higher than in the 1970s. More recent work in Siberia has shown that rates of methane bubbling from thaw lakes are already five times higher than previously assumed. The conclusion is unavoidable: the more frozen land that degrades into stagnant mires, the more methane will be released. And given that permafrost degradation is already accelerating across the Arctic even as I write, this process will be well under way long before global temperatures hit four degrees above today's levels.

Despite the sobering conclusions of these and other studies, that Arctic melting will have a dramatic positive feedback effect on global warming, the extent of this feedback is still unquantified – and is therefore not included in current projections of climate change. But the implication is clear: this dangerous process could be even more significant globally than the carbon cycle changes discussed in the previous chapter. As the Alaskan Arctic specialist Lawson Brigham puts it, thawing permafrost 'is a real wild card in the carbon cycle'.

So how much will the Arctic meltdown add to further global warming? Half a degree? One? More? There is still no clear answer. 'People haven't quite pulled the whole picture together yet,' admits Walter Oechel, an expert on Arctic ecosystems based at the University of California. 'But what we do know is that the potential amounts are huge and very, very scary.' Not everyone agrees that this permafrost time bomb is likely to go off any time soon: one 2007 study argues that the projected melt rate over the century is overstated. But because the amounts of carbon in question are so enormous – perhaps as high as 900 billion tonnes in total – even small changes could have a colossal impact. As Phil Camill says, if just 1 per cent of this potential carbon reservoir were decomposed each year in a warmer world, 'it would be as if we doubled our current rate of emissions'.

This is the ominous conclusion of the four-degree world: that just as with the projected Amazon collapse and carbon cycle feedback of the three-degree world, stabilising global temperatures at four degrees above current levels may not be possible because of carbon releases from Arctic permafrost. In this scenario, if we reach three degrees, that could lead inexorably to four degrees, which in turn could lead inexorably to five. And as we will see in the next chapter, at five degrees an even greater source of methane may come into play. This time the threat comes not from the land, but from the oceans. Once again humanity could be powerless to intervene as runaway global warming continues to push the world into an extreme – and increasingly apocalyptic – greenhouse state.

5°

5

FIVE DEGREES

A new world

With five degrees of global warming, an entirely new planet is coming into being – one largely unrecognisable from the Earth we know today. The remaining ice sheets are eventually eliminated from both poles. Rainforests have already burned up and disappeared. Rising sea levels have inundated coastal cities and are beginning to penetrate far inland into continental interiors. Humans are herded into shrinking zones of habitability by the twin crises of drought and flood. Inland areas see temperatures ten or more degrees higher than now.

With huge quantities of extra heat in the atmosphere, both evaporation and precipitation increase. In the tropics, the convergence zone of the trade winds builds up tremendous downpours, combining with a more intense south Asian monsoon to raise the downstream flows of the Ganges and Brahmaputra rivers by nearly 50 per cent. At the highest latitudes, Siberian, Canadian and Alaskan rivers also increase their flows dramatically due to heavier rains. A resurgent east Asian monsoon dumps nearly a third more water in the Yangtze, and nearly 20 per cent more in the Huang He (Yellow River). The UK experiences almost yearly severe winter flooding.

Together with this increase in rainfall for water-rich regions comes an increase in aridity for the places which already suffer from water shortages. To quote from the conclusions of a scientific modelling team led by the veteran Japanese meteorologist Syukuro Manabe, who explored a simulated five-degree world: 'It is likely that a reduction of soil moisture in semi-arid regions would induce the outward expansion of major deserts of the world, such as the North American desert, the Sahara and Kalahari deserts of Africa, the Patagonian desert of South America, and the Australian desert.' In addition, 'the reduction of soil moisture in the north-eastern region of China could induce the eastward expansion of the Gobi desert' as well.

Although the resolution of the model is too low to get a precise view of predicted changes at the level of individual countries, it does seem to suggest the creation of an entirely new desert area in north-eastern Brazil (the collapsed Amazon), as well as projecting dramatic drying across the southern half of the US (as in the Dust Bowl scenario). In fact, two globe-girdling belts of perennial drought clearly unfold across the map of future five-degree climate, with the northern hemisphere dry belt encompassing all of the central Americas, the entire southern half of Europe, the western Sahel and Ethiopia, southern India, Indo-China, Korea, Japan and the western Pacific.

In the southern hemisphere, an equivalent dry belt grips the southern portions of Chile and Argentina, eastern Africa and Madagascar, almost the whole of Australia and the Pacific islands. Again, a feedback comes into operation: higher evaporation reduces available soil moisture in semi-arid regions, cutting rainfall further and turning them into full-scale deserts. In the worst-affected regions, 40 per cent of available water is lost.

When considered together with the exhaustion of fossil aquifers and the disappearance of snow and glacial melt from mountain chains, these belts imply widening zones of the planetary surface which are no longer suitable for large-scale human habitation. Whereas in the current climate, uninhabitable areas include places like the central Sahara and Gobi deserts, in future these zones will be expanded dramatically. Today huge cities like Cairo and Lima can only be sustained in the middle of deserts by the supply of water from outside via rivers or from underground reserves. Yet, underground aquifers are already unsustainably used, and as we have already seen, surface run-off is projected to decrease further in semi-arid areas. The five-degree model foresees a 20 per cent decrease in Nile flows, whilst Lima's river Rimac will (as we saw earlier) be desiccated by glacial disappearance. The model lacks the detail to pick out changes at small scales in Peru and the US, but according to a separate study, nearly 90 per cent of California's winter mountain snowpack disappears, cutting river flows in the south of the state (where Los Angeles and San Diego are located) by nearly half.

The role of underground aquifers is worth a closer investigation, because they are crucial in supporting cities and agriculture in dry areas. However, as mentioned above, these are by and large a non-renewable or an already overused resource: many of them contain water that fell as rain thousands or even millions of years ago. Yemen, for example, relies for almost all its fresh water on a fossil aquifer which is well on the way to disappearance. Saudi Arabia has recently been getting so desperate that deep wells based on oil-drilling technology have been applied to the task of finding water. In India, China and the United States, water tables are dropping rapidly even in the current climate due to the pressures of population and intensive agriculture outstripping any aquifer recharge coming from precipitation. Both agriculture and, by extension, population will be unsustainable if these water sources dry up just at the same time as new belts of desert are expanding outwards from the tropics.

One solution might be to undertake massive population transfers from evacuated dry regions to newly inhabitable areas in the
far north, particularly Canada and Siberia. As mentioned earlier,
all the major Russian and Canadian rivers see big increases in their
flows, thanks to higher rainfall, and these are also the zones
experiencing the largest temperature increases – dramatically
lengthening growing seasons and reducing the seasonal grip of
winter even in legendarily cold places like Siberia. Could whole
new agricultural breadbaskets be brought into production? Even if
it seems more feasible to grow food remotely than to transfer
whole cities and national populations to these new areas, their
potential as refuges remains clear.

However, before you rush to buy property in Sneznogorsk or
Nizhnevartovsk, take note of the fact that Manabe's model also
projects summertime drying in continental interiors, desiccating
soils which might otherwise be available to grow crops. Areas that
still receive snow in winter may help to smooth the seasonal
change with gradual snowmelt, but major engineering works such
as dams to hold winter rains would become essential to irrigate
any new crops.

It may also become too hot in continental climates – with their
annual extremes of temperature – during the summer months to
grow arable crops. (Siberia, after all, is already hot enough in
summer to see extreme heatwaves and forest fires.) One modelling
study, indeed, looking specifically at global food production,
found that even northern Canada and the former Soviet Union
exhibited agricultural production declines in the five-degree
world. One solution might be to concentrate populations and new
farming colonies on Arctic coasts of Russia and in the Canadian
islands, where the moderating maritime influence should keep
summer temperatures at tolerable levels.

None of this is feasible, of course, if these northern countries decline to take in all these extra refugees – and James Lovelock has even suggested a scenario where China invades Siberia and the US invades Canada to seize the remaining habitable land by military force. Any armed conflict, particularly involving the widespread use of nuclear weapons, would of course have the by-product of further increasing the planetary surface area considered uninhabitable for humans.

As mentioned in previous chapters, recently glaciated soils tend to be thin, rocky and poor, with little in the way of nutrients or organic matter. However, when compared to Africa and Asia – which lose a third of their food supply in the study mentioned above – the higher latitudes would escape relatively lightly. But summer heatwaves in the boreal forest region of Canada would also add to the risk of wildfire outbreaks: one study of fire risk in a tripled-CO_2 climate projects a doubling of the area burned, with 3.8 million hectares reduced to smoke and ashes each year.

Both agricultural expansion and increased wildfires would very likely destroy large areas of boreal forest across subarctic regions of Canada, Alaska, Scandinavia and Russia, putting still more carbon in the atmosphere and accelerating the worldwide loss of biodiversity. But I suspect the survival of the Siberian tiger would be a secondary concern in a world where human survival was itself becoming increasingly precarious. The tiger – and the taiga – would have to go.

Blast from the past

On a chilly summer's day back in 1975, fossil-hunters Mary Dawson and Robert West kept their eyes to the ground as they tramped slowly across the Arctic landscape of Canada's Ellesmere Island. To their east loomed the solid white mass of one of Ellesmere's several ice caps. To their west lay an ice-flecked fjord, its steep sides leading down into the dark blue of the Arctic Ocean. Dawson and West were hoping to discover evidence which might help settle a dispute amongst palaeontologists about whether there was once a land bridge between the high latitudes of Europe, Asia and North America, along which early mammals might have migrated. What they discovered instead was something very surprising about the climate, an accident of fate which dramatically changed today's scientific understanding of the distant past.

Near the head of the fjord, Dawson and West spotted fragments of bone amongst exposed rock outcrops. To their surprise, these were later found to belong to an alligator, a reptile which today only lives in warmer climates thousands of kilometres to the south. Other bones also belonged to subtropical animals: three species of warm-water turtle were unearthed, along with a tortoise and several early mammals. The fossils were dated to the early Eocene, a geological epoch beginning 55 million years ago. The fossil discoveries raised searching questions: what were warm-loving animals doing hundreds of kilometres north of the Arctic Circle? Did they actually live there, or had their bones arrived by some kind of fluke? Geologists already knew that the Canadian landmass was then close to its current position, so shifting tectonic plates couldn't explain the discrepancy. Most importantly, if these subtropical animals had actually lived in the Arctic, how had temperatures even stayed above freezing when the whole region spent several months of the year plunged into polar darkness? No one knew.

Some investigators speculated that perhaps the Earth's tilt away from the Sun was lower during the Eocene, reducing seasonal differences and warming the poles. But no evidence could be found to support this assertion, and obvious seasonal growth rings in unexpectedly discovered fossilised trees showed that there were indeed major differences between seasons. The mystery deepened.

Outside the Arctic, other researchers were gathering evidence of related surprises at the beginning of the Eocene, and the end of the epoch that preceded it, the Palaeocene. Palaeontologists hammering and digging their way through hundreds of metres of sediment in Wyoming's Bighorn Basin discovered dozens of new mammal species, all of which appeared suddenly in North America at around the same time. These included several different types of *Plesiadapis*, a kind of bushy-tailed primate reminiscent of the lemur; various rodents; the chisel-toothed *Esthonyx*; and the fearsome *Coryphodon*, a 300 kg sabre-toothed monster which looked like a cross between a rhinoceros and a bear. All these animals, many of them ancestors of today's mammal species, appeared suddenly at the boundary between the Palaeocene and the Eocene. There was only one likely explanation for their sudden arrival in North America – they had migrated from Asia across the fabled land bridge in the high Arctic. Given that the ice and freezing temperatures would stop any such migration today, this suggested a much warmer Arctic in the distant past.

Out at sea, something strange was also going on at the end of the Palaeocene. Whatever it was, life turned pretty unpleasant for organisms living on the ocean bed at the time. Whereas normal sea floors see continual churning from worms, clams and tiny single-celled creatures called foraminifera, two Californian geologists examining an ocean core from the Weddell Sea off Antarctica in 1991 were surprised to find very little disturbance in 10 centimetres of the muddy ooze they were looking at – 10 centimetres which just happened to coincide with the same Palaeocene–Eocene transition as the fossil crocodiles from Canada. The two scientists, James Kennett and Lowell Stott, quickly realised that their section of undisturbed mud showed that almost everything which normally lives in and on the sea floor had suddenly disappeared. Kennett and Stott had stumbled across a major deep-sea extinction event, perhaps the largest for tens of millions of years. They were nearly certain that the extinction was climatic in cause: a sudden warming of these ancient Antarctic waters had deprived bottom-dwelling organisms of oxygen. The deep ocean had turned anoxic, poisonous to oxygen-breathing life. So not only had the polar atmosphere in the Arctic got much warmer, but here was evidence that the seas – even in the depths of the Antarctic Ocean – had suddenly heated up too.

But this discovery merely raised further questions. None of these studies had given any hint of what might have caused the sudden warming of the Palaeocene climate. Unlike at the earlier Cretaceous–Tertiary boundary – when the dinosaurs were wiped out by an asteroid – there is no evidence of a catastrophic extra-terrestrial impact at the end of the Palaeocene. Yet in geological terms the event was nearly instantaneous. Why had the climate suddenly flipped?

Enter Gerald Dickens, a palaeoceanographer then at the University of Michigan. Dickens's work focused on an unusual substance called methane hydrate, an icelike combination of methane and water that forms under the intense cold and pressure of the deep sea. Dickens was not at the time trying to explain the past: he was trying to help oil companies discover whether methane hydrates – which could potentially double the world's energy reserves – could be drilled to produce natural gas. It was occasionally an exciting venture: on one expedition, his drilling tube dramatically exploded due to the pressure of methane within, firing mud fifty metres into the air. Luckily nobody was injured.

As Dickens would later explain, the important thing about methane hydrate is that it is only stable when kept very cold or under high pressure. That is why his drilling tube had blown: as it emerged from the ocean surface, warming up and losing pressure, the methane within had turned from ice to gas in an explosive chain reaction. Methane is also an important greenhouse gas, molecule for molecule twenty times more powerful than carbon dioxide. With his knowledge about the unsolved questions of past climate change, Dickens hypothesised that perhaps methane hydrate releases might have happened in the past planetwide. That would have pumped large amounts of greenhouse gas into the atmosphere, heating up the global oceans and releasing yet more methane hydrate in a comparably explosive chain reaction. Could this explain the sudden warming of 55 million years ago? Dickens thought it could. It was, he suggested, as if the oceans had let rip a giant belch of methane, pushing global temperatures through the roof.

But he still lacked hard evidence – until in 1999 fellow palae-oceanographer Miriam Katz seemed to find what everyone was looking for. At 512 metres down another ocean core, this one taken off the coast of Florida, Katz found evidence of a submarine landslide, closely followed by an isotopic methane signature in the sediment and the now familiar mass extinction of sea-floor organisms. Katz's ocean core suggested that massive avalanches had tumbled down the continental shelves, explosively releasing methane hydrates in enormous quantities. Here was a possible smoking gun: direct evidence of a 55-million-year-old gas blast which had perhaps helped catapult the Earth into an extreme greenhouse state.

Methane hydrates may not have acted alone: research published in the journal *Science* in April 2007 fingers a different culprit altogether. This particular agent of change left clues scattered right across the North Atlantic. You can find them in the Faroe Islands, under the sea near Rockall, on the eastern side of Greenland, and even in Northern Ireland – basaltic rocks which tell the story of a monumental set of volcanic eruptions. These were not explosive blasts like those of Mount St Helens or Krakatoa, but long-term outpourings of basaltic lava into layers which in eastern Greenland reach a staggering 5 kilometres thick, covering in total 1.3 million square kilometres of the modern-day North Atlantic sea floor. This magma didn't intrude just anywhere: it pushed up into carbon-rich coal-bearing sediments, heating them up and releasing vast quantities of methane and carbon dioxide in the process. Geologists have dated the point when the most magma erupted as 55–56 million years ago – the same time as the hot period at the end of the Palaeocene.

That this enormous and long-lasting volcanic episode – unique in the geological record for tens of millions of years – took place at almost exactly the same time as an era of massive global warming is unlikely to be just coincidence, particularly as similarly massive volcanic eruptions are linked with similarly extreme greenhouse episodes even longer ago in the Earth's history – as the next chapter will reveal. Moreover, the million-year-long eruption may have had another greenhouse-related effect too, by uplifting sea-floor sediments and thereby reducing the water pressure on methane hydrates trapped in the sediments. This too could have led to a catastrophic gas release, possibly containing 2,800 billion tonnes of carbon: more than enough to account for the dramatic change in the climate.

Whatever the source, the impact was as profound as it was global. Summer heatwaves scorched the vegetation out of continental Spain, leaving a desert terrain which was heavily eroded by winter rainstorms. Rubbly deposits on the southern slope of the Pyrenees attest to the fearsome nature of these monsoonal flash floods – one 'megafan' reported by geologists in 2007 covers an estimated two thousand square kilometres, and was formed in just a few thousand years as rising atmospheric CO_2 levels flipped the Earth into a new greenhouse state. In the oceans, this CO_2 rapidly dissolved, turning the seas acidic. In North America (modern-day Utah), there was a similar increase in seasonal extremes, with monsoonal storms lashing the soil. This new monsoon also brought abundant rains to the slopes of the Rockies, supporting tropical vegetation in an area which is now cold and semi-arid. Palm mangroves grew as far north as England and Belgium.

In the high Arctic, rainforests of dawn redwood (*Metasequoia*) appeared: on Axel Heiberg Island in Canada, stumps of mummified trees are still visible today, so well preserved that their wood will still burn, tens of millions of years after they died. These trees flourished in temperate conditions, creating their own greenhouse effect by releasing water vapour to insulate the region during the dark polar winters. Despite the darkness, the Arctic climate was nothing short of subtropical: sediment cores drilled out from the central Arctic Ocean show that sea temperatures close to the North Pole rose as high as 23°C, warmer than much of the Mediterranean is today. Air temperatures may even have reached a staggering 25°C, again indicating a subtropical climate with no frost or snow during any part of the year. The whole area experienced much higher rainfall as storms tracked northwards – leaving the true subtropics and mid-latitudes in contrast stricken by drought. As you might expect from the temperature, the polar Arctic Ocean remained entirely unfrozen: indeed, no ice at all was to form there for another 15 million years.

This was a world where carbon dioxide in the atmosphere reached dangerously high levels, and average temperatures soared by five degrees Celsius. It was a world with acidic oceans, rapidly changing ecosystems, ice-free poles and extremes of wet and dry. In short, it was a world much like the one we are heading into this century.

Many scientists have already recognised this similarity, which is why the event now known as the Palaeocene–Eocene Thermal

Maximum (PETM) has attracted such widespread attention from the scientific community. Indeed, almost every academic study published about the PETM in recent years mentions that it can be seen as a natural version of whatever human-caused global warming might have in store. One of the first to recognise the importance of the PETM as a 'natural analogue' for current greenhouse gas releases was Gerald Dickens, who wrote in *Nature* as early as 1999 that 'we can now begin to view aspects of Earth's future in an entirely new light'. In May 2006, the Harvard University scientists John Higgins and Daniel Schrag reaffirmed this view, stating: 'The PETM represents one of the best natural analogues in the geologic record to the current rise in atmospheric CO_2 due to burning of fossil fuel.'

Although the *total* carbon input into the atmosphere 55 million years ago was larger than humans have so far managed – with CO_2 levels of more than 1,000 parts per million persisting into the early Eocene – the *rate* of greenhouse gas addition is actually faster now than then. The palaeoceanographer Jim Zachos told the 2006 meeting of the American Association for the Advancement of Science that today's human carbon emissions are perhaps 30 times faster even than the massive postulated methane belch of the PETM. And judging by carbon isotope ratios in rocks spanning the Palaeocene–Eocene boundary, we are already about halfway to the kind of searing global heatwave that was experienced then by life on Earth.

The likely role of methane hydrates in causing this heatwave also offers another worrying lesson for humanity. Vast amounts of the same methane hydrates still sit, quietly biding their time, on subsea continental shelves around the world. With the oceans now warming up, there is a chance that some of this hydrate will be destabilised and vent catastrophically into the atmosphere in a terrifying echo of the methane belch of 55 million years ago. This would boost atmospheric temperatures further, adding to an unstoppable feedback of runaway global warming. Humans would sit powerless to intervene as their planet began to turn into Venus.

But just how likely is this apocalyptic scenario? Unfortunately, few scientists have hazarded a guess so far. It isn't even clear just how much methane hydrate there is out there, or how quickly it might respond to ocean warming. Reassuringly, perhaps, it has been suggested that it took about ten thousand years for the methane hydrate release to fully play out during the PETM. That might be instantaneous to a geologist, but on a human timescale it seems much less scary. Remember: it takes centuries for higher temperatures at the top of the ocean to propagate down into the depths, suggesting that hydrates are unlikely to destabilise and collapse all at once. 'Runaway' global warming could still happen, but the positive feedback cycle would take thousands of years to complete.

However, as we saw above, carbon is being added to the atmosphere much faster now than during the PETM, raising the stakes somewhat. If warming is extreme enough at the surface to reach the bottom of the ocean more quickly, we could be in for a rough ride. A modelling study by two methane hydrate experts, Bruce Buffett and David Archer, suggests that the store of hydrates on the ocean floor would decrease by 85 per cent in response to a warming of just three degrees – but they don't say how long this might take.

Writing elsewhere, David Archer has suggested that the Arctic Ocean is the one to watch: it is relatively shallow and likely to see the most extreme warming, so the methane hydrate response time there could be much faster. The UK's Hadley Centre reinforces this concern with a map which suggests that large areas of the Arctic Ocean which may currently support methane hydrates will be in the zone of total melt as early as 2090. Although no one is sure exactly how much of the stuff is sitting there on the floor of the Arctic Ocean waiting to melt and explode up to the surface, even modest amounts would be enough to give a significant boost to global warming.

Tsunami warning

We are accustomed to thinking that if we take our collective foot off the carbon accelerator then global temperature rise will begin to slow. That is currently true. But if substantial methane hydrate melt begins to occur in the Arctic Ocean basin, then the accelerator will be jammed, and there will be nothing we can do to cut the speed of climate change. Again, no one can say for sure where this tipping point might lie, but it stands to reason that the harder we push the climate, the closer we are likely to get to the edge of this particular cliff.

Buffett and Archer conclude their paper, somewhat enigmatically, with the following statement: 'It is not known if future warming is sufficient to cause failure of continental slopes on a global scale, but isotopic evidence of rapid carbon release in the past is suggestive.' In other words, Miriam Katz's evidence of catastrophic subsea landslides could also offer a grim warning for the future, one where large-scale methane hydrate releases destabilise the sloping ocean bottom and spark similar massive avalanches.

Like shifting tectonic plates, these oceanic landslides can displace huge quantities of water. When this happens, strong shockwaves propagate outwards from the area of disturbance. What these waves look like became suddenly clear to the whole world on 26 December 2004, when the tectonic plates off the western coast of Indonesia shifted dramatically in a magnitude-9.2 earthquake. Everyone knows the name of these waves: they are called tsunamis.

Way back in 1965, two Scottish geologists stumbled across evidence that just such a tsunami hit the British Isles over 8,000 years ago. Walking through the western part of the Forth valley, not far from Edinburgh, they discovered a thin layer of sand in amongst the more common peat. The sand layer was clearly deposited by a major event, as it continued for over a kilometre – the two geologists suggested a big flood in the river might have been the cause. But later investigations at other sites in eastern Scotland turned up similar deposits, which even extended as far south as the coast of Northumberland. The deposits were most extreme in the Shetlands, where churned-up mud and boulders were discovered in peat bogs many metres above the high tide mark. Clearly, localised events like river floods or storm surges could not have been to blame. The discovery of a major submarine landslide site off the coast of Norway in the late 1980s offered conclusive proof: a massive tsunami had hit the UK.

The Storegga Slide, as the submarine avalanche later became known, was truly gigantic, shifting 3,500 cubic kilometres of sediment downslope from the Norwegian continental shelf into the deeper Arctic Ocean. The resulting tsunami inundated over 600 kilometres of coastline around the North Sea, with run-ups of 3–6 metres above sea level in eastern Scotland, 9–12 metres in western Norway, more than 10 metres on the Faroe Islands, and a devastating 20 metres in the Shetlands – comparable in impact to the 2004 Asian tsunami strike in Banda Aceh. Like that tragic tsunami, most scientists think that the Storegga Slide was caused by an earthquake – though the sudden release of methane hydrates may also have been a factor. Tellingly, the sea floor is covered with pockmarks where huge quantities of gas once erupted. In addition, whilst unaffected sea-floor sediments next to the Storegga Slide are still rich in methane, the avalanched portion has lost any methane hydrate it may once have had.

This wasn't enough to change the climate at the time, but it does suggest that methane hydrate destabilisation and submarine landslips go hand in hand. And unfortunately, if one of these things happens, the first warning most coastal dwellers have will be towering waves racing towards the shore.

The prospect for humanity

In some respects, despite the tsunamis, the world at the beginning of the Eocene sounds quite pleasant. Without chilly ice caps to cool things down, lush forests grew right up to the poles. Places which would normally experience a temperate climate became subtropical, and a fascinating array of species spread across the globe.

But don't be deceived. The world in a natural state can never be a perfect analogue for the globe as it exists now. We are already well into a new geological era, the Anthropocene, where human interference is the dominant factor in nearly every planetary ecosystem, to the detriment of perhaps all of them. The rising heat of the Palaeocene–Eocene Thermal Maximum took place over about 10,000 years, giving plants and animals time to migrate and adapt to the new circumstances. Geological evidence from North America shows that subtropical plant species were able to shift their ranges 1,500 km north in that time, from what is now Mississippi to Wyoming. Even so, as we discussed earlier, many less adaptable species lost the battle and were wiped out.

We don't have ten thousand years. The changes described here would be in place a few decades from now – a pace of warming much too rapid for substantial adaptation either by natural ecosystems or human civilisation. This may indeed, as suggested earlier, be the fastest large-scale climate warming the world has ever experienced – faster even than climate changes which caused catastrophic mass extinctions, as the following chapter will show. Particularly deceptive may be the impression that the PETM hot spell was well watered, with heavy monsoon rains quenching the thirst of places which are now semi-arid. In fact the first plant migrants to arrive in Wyoming were small-leaved and drought-tolerant, suggesting that the PETM began dry as well as hot.

This supports the modelling evidence discussed earlier about spreading deserts and water shortages; monsoons such as those which characterise the later part of the PETM would take thousands of years to evolve. During the transition phase, higher temperatures would evaporate more water, leaving already arid areas even drier than before. This is why a resurgent Chinese monsoon is unlikely to give any imminent relief to the drought-stricken north of the country, and why Pakistan's failing glacial rivers will not be replenished any time soon by a stronger Indian monsoon. It also supports by implication projections of a much more arid climate in the northern half of South America, with new areas of desert spreading west as the Amazonian ecosystem collapses.

Earlier we touched on the concept of 'zones of uninhabitability': places where large-scale, developed human society would no longer be sustainable in the five-degree world. Looking at the geological evidence of dramatic changes at the start of the Eocene, however, it is clear that even this discussion may be overly optimistic. Instead, we perhaps need to start talking about zones of inhabitability: refuges.

With the tropics too hot to support most food crops and the subtropics out of production due to perennial drought, the region where large-scale human civilisation remains feasible – the 'belt of habitability' – contracts towards the poles. (It is worth remembering that this applies as much to the seas as to the land: the destruction of coral reefs and rapid warming of the oceans will

209

likely wipe out most marine life within the same tropical and subtropical belt.)

For humanity, a new era of enforced localism is likely, where globalisation goes into reverse and people reassert more restricted identities. Our economy is globally interconnected at present, with huge volumes of trade taking place between far-flung regions. But hypothetical customers in some ravaged coastal city of the future will no longer be able to buy, whilst producers in a drought-stricken subtropical zone will no longer have anything to sell. Well before this situation is reached, the sensitive and volatile capital markets will surely have collapsed, erasing ownership bonds between foreign and domestic capital, and precipitating a world-wide economic depression. The Great Depression of the 1920s and 30s showed how difficult it is for societies to adapt to such pressures, and also how unpleasant political philosophies can gain traction as social insecurities rise.

As suggested earlier, powerful civilisations, when confronted with the collapse of their habitable homelands, may seek to shift their populations into subarctic areas in order to stave off wide-spread starvation and internal conflict. An analogy might be Hitler's concept of *Lebensraum*, an empire carved out of Russia and eastern Europe to cater for the population of an expanding Third Reich. James Lovelock has also suggested that Africa – the birthplace of our species – may be able to support small refuges in rainy highland regions. Candidates might include areas like the Ethiopian highlands and Lesotho in South Africa, mountain fast-nesses where agriculture can persist, and isolated valleys are more easily defended from marauding intruders. Needless to say, the era of food aid and international assistance would be long gone.

Elsewhere in the world, I imagine that northern Europe, includ-ing the British Isles and Scandinavia, could become a similarly crowded and contested refuge. The region seems likely to remain within the winter-rainfall belt, although the collapse of the Gulf Stream – northern Europe's wild card – could leave Britain in a drier climate, with temperatures stabilising or perhaps even cooling for a time. In the southern hemisphere, westerly winds should continue to deliver plentiful rainfall to Patagonia and Tierra del Fuego in Chile, whilst even further to the south, human colonies might also survive on the newly ice-free Antarctic Peninsula, perhaps moving slowly south as the retreating ice makes new land available. Tasmania and New Zealand's South Island also remain within the temperate rainfall belt, and might also offer refuge to survivors from hotter regions further north, such as Australia and Indonesia – though they of course lack the land area to be much help to climate refugees who by then may well number in the hundreds of millions.

In all these cases, migrants would be well advised to establish their new communities at a safe distance from the coast. Sea level rise will already have turned many coastal cities into wave-battered ghost towns, their abandoned buildings crumbling into beach sand with each new storm. Rising waters will continue to inundate the land for centuries to come, driving out agriculture from low-lying plains like the Nile, Yangtze and Meghna (Ganges–Brahmaputra) deltas, as well as from fertile flatlands like those along Britain's eastern shore.

Coastal residents will also need to keep a wary eye out to sea for more sudden threats. Far offshore, methane hydrate deposits down on the seabed may be bubbling into life, their deadly cargo of gas surging to the surface in explosive blasts. To any observer, it would look as if the sea was boiling, perhaps with faint yellow flames dancing above the churning surface as some of the methane burns off. Far underneath, a shudder in the water and a rumble from the seabed would attest to the shifting of millions of tonnes of sediment in a submarine landslide. The clock would be ticking, and a tsunami – barely noticeable out at sea, but rearing up to stupendous heights nearer the shore – would have been unleashed. We know what the warning signs would be: a gradual retreat of the water away from the shore, then a white line on the horizon, a wall of water racing towards the beach, followed by a deadly debris flow of mud and branches flattening everything for several kilometres inland. Life would be precarious anywhere that is not securely located tens of metres above the restless seas.

Survival

Where no refuge is available, and crops and water supplies fail, civil war and a collapse into race or community conflicts seems – sadly – the most likely outcome. By and large history teaches us that humans do not sit and starve in situ when times get bad – they take what weapons they can find and move to more promising regions, triggering warfare with whatever groups already inhabit the contested area. Our tribal inheritance also mentally precondi-tions us to blame 'outsiders' for perceived injustices or shortages, just as the Jews of Europe were once persecuted for supposedly hoarding food in times of famine. Conflicts which were once fought with spears and swords, however, will now be fought with guns, grenades or nuclear weapons.

So how might one go about planning to survive? Most people's natural response would probably be to stake themselves out an isolated patch of mountain where they and their loved ones might lay low until the crisis passes. This could indeed be an option in regions with large landmasses and sparsely populated highlands, like the western United States. Some places like Montana already have a strong survivalist tradition, though not, needless to say, because of concerns about global warming.

In reality, however, relatively few people have this option. How many of us could really trap or kill enough game to feed a family? When the closest most city dwellers come to exercising hunter-gathering skills is jumping the checkout queue at the supermarket, how many modern humans would really be able to live off the land? Even if large numbers of people did successfully manage to fan out into the countryside, wildlife populations would quickly dwindle under the sudden pressure of human predation. Supporting a hunter-gathering lifestyle takes ten to a hundred times the land area per person that a settled agricultural community needs. A large-scale resort to survivalism would turn into a further disaster for biodiversity as hungry humans killed and ate anything that moved, much as the bushmeat trade has devastated wildlife in tropical Africa today.

In more densely populated regions, like Europe or China, isolationist survivalism is simply not an option. Nowhere is properly remote and defensible whilst simultaneously offering enough resources for survival. Another option might be stockpiling: hiding stashes of food and drinking water and trying to sit out the collapse. But defending your supplies from hungry invaders is never easy; over a long period it is next to impossible. Hunger is a powerful motivator, and people maddened by hunger or jealousy do not give up. Sooner or later you run out of ammunition, or get caught asleep with your defences lowered. Looters who are truly desperate – as the hurricane disaster in New Orleans briefly showed – soon lose their fear of guns.

In a situation of serious conflict, invaders do not take kindly to residents denying them food: if a stockpile is discovered, the householder and his family – history suggests – may be tortured and killed, both for revenge and as a lesson to others. Look for comparison to the experience of present-day Somalia, Sudan or Burundi, where conflicts over scarce land and food are at the root of lingering tribal wars and state collapse. Most of human history is full of such dark episodes of genocide, rape and plunder: our relatively prosperous interlude may prove to have been a lucky aberration, thanks in large part to the massive boost in food and energy that our civilisation derived from fossil fuels. This same fossil energy boost, of course, while allowing our species to proliferate massively in numbers and construct wonderful, complex societies in little more than a historical instant – could in the longer term prove to be our undoing.

A drastic reduction in human populations is unambiguously the most likely outcome of a rise in global temperatures towards five degrees – what James Lovelock unhappily terms 'the cull'. Even at present numbers, the planet will have trouble supporting human society indefinitely, as we already see in a myriad of ways from overfishing to soil erosion. But with human population growth projected to add still further to our ballooning numbers, the overall situation will become steadily more precarious as the world warms up. I find it difficult to avoid the conclusion that millions, and later billions, of people will die in such a scenario. In Gaian terms, I suppose, the planet would be trying to restore a balance.

Unbelievably, perhaps, this still isn't quite the worst-case scenario. The next chapter will show how humanity's survival, even as a species, could be threatened by the ultimate apocalypse: six degrees of global warming.

6°

6

SIX DEGREES

As we enter a world six degrees warmer than today's, there are few clues to what really lies in store. My Virgil guides in this latter-day version of Dante's Inferno have so far been mostly climate modelling scientists, yet the majority of them have fallen by the wayside now: the current generation of climate models almost all stop short of simulating six degrees of warming by 2100. (But as we have already seen, models do have the tendency to be conservative by design, so this outcome cannot be discounted – and indeed is part of the IPCC's scenario of projections on which this book is based.) Instead, we must rely on sketchy geological information about extreme greenhouse episodes in the Earth's distant past to light our way forwards into this Sixth Circle of Hell. Dante offers a warning to his readers, and so shall I: were this to be a television programme, it would be prefaced by a note of caution – some viewers may find some of the following scenes upsetting.

The Cretaceous world

The longest-lasting extreme greenhouse episode – the Cretaceous period – was for most of the time a relatively benign one, albeit on a planet which was geographically and ecologically very different from the one we know today. During this period, between 144 and 65 million years ago, ferns, cycads and conifers dominated the land; flowering plants were only just beginning to evolve. The great supercontinent Pangaea was tearing down the middle, splitting South America from Africa like pieces on a giant floating jigsaw. The narrow corridor of water between them, the young Atlantic, was no wider than the Mediterranean is today. As the tectonic plates shifted by a few millimetres each year, huge volcanic eruptions shook the planet.

In the southern hemisphere, India was far to the south of its current position, still drifting peacefully off the eastern coast of Madagascar. The major continents also looked very different: with sea levels 200 metres or more higher than today's, many continental interiors were flooded by the oceans. North America was split into three distinct islands by the oceanic invasion, whilst parts of North Africa, Europe and South America also disappeared beneath shallow seas. These marine incursions left distinctive limestone platforms, still visible today from the Mediterranean to China. They also left chalk: indeed the Latin word for chalk, *creta*, is the origin of the name 'Cretaceous'. England's famous white cliffs and chalk downlands all date back to the Cretaceous period.

The world was also a much flatter place. Mountains form when plates collide, but the Cretaceous continents were tearing apart, not crashing together. With higher sea levels and smaller continents, only 80 per cent of the current land area existed – the rest was deep blue sea. These geographical differences are as profound as the climatic ones, for during the Cretaceous global average temperatures were ten to fifteen degrees above today's levels – not just for a brief interval, but for millions of years.

Signs of this long-lasting extreme greenhouse climate are evident around the world in rocks laid down at the time. Fossil tree trunks, looking very much like modern palm trees, occasionally emerge from the frozen sediments of Alaska's North Slope. Dinosaurs – some, like the herbivorous, duck-billed *Edmontosaurus*, nearly 20 metres long – grazed in these lush sub-polar forests, leaving bones, footprints and even skin impressions in the Cretaceous rocks. Frosts were either rare or unknown, even at the edge of the Arctic Ocean. The Siberian north-east of Russia basked in year-round Mediterranean temperatures despite the two-month period of polar darkness. Ancestors of the crocodiles – the aptly named champsosaurs – swam in the shallow, warm marshes of the high Canadian Arctic, stalking shoals of passing fish. Groves of tropical breadfruit trees flourished on the west coast of Greenland.

But this world wasn't all balmy sunshine, grazing dinosaurs and gently waving palms. Some rock formations show aggregated deposits called 'tempestites', formed by the rubble from intense storms. These ferocious hurricanes – much stronger than today's, thanks to the hotter oceans – even left their mark on the ocean floor, building up big hummocks which have since been studied by geologists. During the mid-Cretaceous, when carbon dioxide and temperatures were highest and the greenhouse at its peak, these storm-wave hummocks are the largest of all.

As these tempestite deposits suggest, a more intense hydrological cycle drove much heavier rains in some areas. In the flooded interior of North America, which enjoyed a tropical climate, rainfall rates were as high as 4,000 mm per year, drenching the land with the kind of downpours that are today experienced in monsoonal India. Ocean temperatures driving these rainstorms were far higher than today: in the tropical Atlantic they may have reached as high as 42°C, more akin to a hot tub than an ocean. In the sub-polar South Atlantic, near the Falkland Islands, sea surface temperatures commonly averaged 32°C, hotter than in most of the deep tropics today.

Assessing all the geological evidence into one big picture, distinct zones soon emerge. Around the equator a broad humid belt would have seen the heaviest rains and most ferocious storms – but supported few coral reefs and almost no rainforests. A much broader arid area, featuring only drought-tolerant plants and animals, encompassed the rest of the tropics and subtropics, including all of Africa, South America and the southern parts of the US and Europe.

The higher mid-latitudes were warm and humid, but subject to frequent intense burning: some Cretaceous species of fern had the same adaptation to fire that eucalypts do in today's Australia. Plant physiology was also adapted to drought: fossil trees from southern England show uneven growth rings from arid years when the rains failed to arrive. In the polar regions, a humid and temperate climate supported forests in both hemispheres – Siberia had luxuriant growth, as did the Antarctic Peninsula. Needless to say, it was a world without ice caps on either pole. Evergreen forests may even have grown at the South Pole itself (where they would have spent nearly half the year in polar darkness), although today 3 kilometres of ice prevent the discovery of any fossil wood to prove it. At the North Pole, ocean temperatures may have reached a balmy 20°C.

Overall, carbon dioxide levels are thought to have been somewhere between three and six times today's levels, though the two are not directly comparable because greenhouse warming in the Cretaceous would have been offset by a slightly fainter sun. Most of this extra CO_2 was volcanic, thanks to higher amounts of volcanism related to the splitting up of the Pangaean supercontinent. Whereas today volcanoes account for only 2 per cent of the atmospheric carbon dioxide input each year, the eruptions during the Cretaceous were truly massive in scale and persisted for many thousands of years.

But the Earth system always strives for a balance, just as do warm-blooded animals like humans which unconsciously alter their metabolism to maintain a constantly optimum body temperature. Indeed, this view of the planet as a self-regulating organism is a central tenet of James Lovelock's Gaia theory. Lovelock stops short of suggesting that the Earth is a sentient being, yet his observation that various planetary mechanisms act almost in an intentional way to maintain a temperature favourable to life is an accurate one. The operation of the long-term carbon cycle illustrates this particularly well: if carbon dioxide levels in the atmosphere rise too high, then life will be endangered because of an accelerating greenhouse effect – that is why Venus is a dead planet. But if they drop too low, the planet will freeze over. Only a relatively narrow band of carbon fluctuations is desirable; therefore living mechanisms tend to release carbon if it falls too low, and absorb it if it rises too high.

The biggest of these living carbon 'sinks' during the Cretaceous were the huge marine calcium carbonate platforms in the subtropics, composed of layer upon layer of shells. Some of these platforms, which covered tens of millions of square kilometres of shallow sea floor, were later exposed as limestone pavements in places like Mallorca and Greece – indeed, you can often see agglomerations of ground-up shells if you look carefully at the limestone rocks. But the construction process was slow: it took a million years to build up every 30 metres of limestone.

Also important in sequestering carbon, however, was vegetation. Big domes of peat built up beneath forests and within swamps, and were gradually compressed into coal. The fossil forests of the Alaskan North Slope include thick coal layers, whilst Cretaceous coal is also found in north-eastern Russia, western Canada, the interior United States, Germany (much of it as dirty 'brown coal', or lignite), northern China, Australia and

New Zealand. Significant coal deposits may also sit under the ice cap in Antarctica, testament to a warmer period when the polar continent still supported major forests.

Large amounts of carbon were also trapped in ocean sediments as the decaying remains of plankton settled on the ocean floor into layers of rich organic mud. Some of this carbon, since 'cooked' by geological processes and squeezed through pores in the rocks into reservoirs, is a very familiar substance to modern humans: oil.

An obvious lesson should be learned from the working of this ancient carbon cycle. Life on Earth laboured for millions of years to remove dangerously high levels of carbon dioxide from the ancient atmosphere and so keep global temperatures within tolerable limits. Much of this is the same carbon that humans are now labouring to return to the atmosphere through the burning of coal, oil and gas for energy. (They are not called 'fossil fuels' for nothing.) Moreover, humans are a good deal more efficient at shifting carbon than mussels, oysters and plankton – we are releasing it about a million times quicker than Cretaceous life forms managed to sequester it all those eons ago.

Nevertheless, with its lush, coal-forming forests and flourishing animal life one might be lulled by the geological evidence into a sense of the Cretaceous as quite an attractive place, if rather hot and sticky. After all, does it not indicate that the Earth can survive – indeed, that life can flourish – with much hotter global temperatures? Might this not assuage some of our worries about the future? Perhaps. But the Cretaceous ecosystems evolved in the greenhouse climate over a very long period, and many of the plants and animals now turning up as fossils were clearly supremely adapted to it. This is not the case today: we share the planet with species which are largely adapted to cooler conditions. If we do succeed in tipping the Earth back into the extreme greenhouse climate of the Cretaceous, few of the ecosystems we know would survive. The pump is primed, as we'll see later, not for flourishing palm trees in Alaska, but for the worst of all earthly outcomes: mass extinction.

Oily oceans

A more applicable analogue from the past might not be the sustained greenhouse of the Cretaceous, but some of the more rapid global warming events that have hit the planet over the ages. The Palaeocene–Eocene Thermal Maximum, discussed in the previous chapter, was one. In the Cretaceous, too, there were similar temperature spikes, and they were similarly associated with dramatic changes in global climate and living things. These warming spikes are partly responsible for the abundance of crude oil in the Earth's sediments – one global warming spike ironically laying the foundations for another.

The signs of these warming spikes are strips of black shale amongst the otherwise calcareous Cretaceous rocks – the remains of stinky muds that were laid down by a rain of plankton and other marine organisms onto the ocean floor. Under normal conditions, this organic carbon would have been consumed by sea-floor creatures. But during the warming spikes, something went wrong in the oceans: oxygen levels declined, turning them gradually anoxic (stripped of oxygen). With bottom-dwelling creatures driven out, the seas would have resembled stagnant ponds, with only a thin layer of life in the uppermost oxygenated sections. No one is sure what caused these so-called 'ocean anoxic events', or exactly how they progressed – but their correlation with warming peaks seems clear.

One suggestion is that catastrophic methane hydrate releases warmed the climate so severely that the oceans ceased to turn over properly. In the atmosphere, warming drives convection because it happens from the bottom up – warmer air expands, becomes lighter and rises, and the air circulates as a result. In the ocean, warming is top-down; so the lighter warm layer stays as a lid on the colder layers below, shutting off the oxygen supply and potentially leading to massive die-offs. Sudden ocean layering happens during El Niño years off the coast of Peru when the warm current arrives, decimating fish stocks and seabird colonies which normally thrive on the cold upwelling Humboldt current.

This stratification of the oceans also explains why tropical waters look clear and pristine: they are so low in nutrients that almost nothing can survive in them, making them marine deserts. In contrast, the well-mixed colder oceans closer to the poles support great blooms of plankton, making the seas look murky and green but producing abundant fisheries. During ancient 'ocean anoxic events' this stratification was not limited to a particular region, but struck the oceans on a global scale, causing mass extinctions of marine life.

A different hypothesis is that these extreme greenhouse episodes saw a more rapid hydrological cycle, with intense rainstorms washing nutrients off the land surface, sparking a worldwide algal bloom. A modern version of this might be the poisonous 'red tides' that wash up off the coast of China every year, or the anoxic 'dead zone' in the Gulf of Mexico caused by agricultural pollutants streaming down the Mississippi. Stronger winds may have also blown nutrient-laden dust out to sea, just as sandstorms from the Sahara fertilise the Atlantic today; especially since deserts in the Cretaceous would have been much larger.

A particularly dramatic theory has been advanced to explain the biggest warming spike and ocean anoxic event of all, 183 million years ago in the Jurassic. This episode saw carbon dioxide concentrations in the atmosphere leap by 1,000 ppm, pushing up global temperatures by about six degrees in a striking parallel of the IPCC's projected worst-case scenario. The impacts were profound, resulting in the most severe marine mass extinction of the entire Jurassic and Cretaceous periods (a span of time 140 million years long). Geologists dispute the possible causes of the event: one theory suggests that hot volcanic magma intruded into ancient coal seams across thousands of kilometres of southern Africa. In a similar episode to that which may have unfolded at the end of the Palaeocene (as examined in the previous chapter), hot lavas could have gasified the coal, pouring methane and carbon dioxide into the air, and triggering an accelerated global warming spike which stripped the oceans of oxygen.

Strikingly, geologists have discovered thousands of vertical rock pipes, each between 20 and 150 metres in diameter, through which about 1,800 gigatonnes of CO_2 would have vented into the Jurassic atmosphere from these volcanically baked sediments. These 'breccia pipes' are dotted all over South Africa's Karoo plain, and look from aerial photos like tiny volcanoes – a faint geological echo of modern power station chimneys, which of course perform a similar carbon-releasing role today.

Another theory involves that usual suspect: methane hydrates from the ocean shelves, which might have released a pulse of as much as 9,000 billion tonnes of the gas from under the seas. Perhaps some combination of the two explains the dramatic warming. Either way, the entire geological carbon cycle somehow short-circuited, blowing the Earth's climatic fuse – a warning from the distant past, but about our very immediate future.

The effect was devastating, but most species alive at the time did manage to struggle through the crisis, perhaps because it happened relatively slowly. The same cannot be said for a similar but much worse disaster which befell both marine and land-based species at the end of the Permian period, 251 million years ago. That episode was the worst crisis ever endured by life on Earth, the closest that this planet has come to losing its wonderful living biosphere entirely and ending up a dead and desolate rock in space. If the postulated Jurassic coals event was a blown fuse, the end-Permian mass extinction was more akin to the whole house burning down.

The end-Permian wipeout

To the Chinese quarrymen hacking away at the rocks of the Meishan quarries in the Zhejiang province of south China, the boundary between grey limestone and darker mudstone must have looked thoroughly unremarkable. They might have noticed that the rock just below the boundary was more crumbly than normal, making it annoyingly unsuitable for building. They might also have spotted the sudden change in colour from pale grey to near-black in the layers of stone. But blasting and hauling away these drab-looking rocks would just have been another day's work. None of the workmen would have realised that, with their drills, picks and shovels, they were in the process of uncovering one of the most important geological sections ever discovered. They had sited their quarry right on the Permian–Triassic boundary, scene of the worst mass extinction of all time.

The Meishan sections have become the geological gold standard for the end-Permian because the succession of rock beds is so clearly defined. They were laid down on a shallow seabed, and the limestone sections below the Permian–Triassic boundary are stuffed with fossils. Particularly abundant are tiny micro-organisms called foraminiferans and conodonts, but sea urchins, starfish and small crustaceans are also found, together with corals, fishes and sharks. Clearly the sea before the extinction was highly productive and full of life, each animal and plant well adapted to its evolved place in a complex web of ecosystems.

Then disaster strikes. The fossils disappear, and the limestone is replaced by a churned-up layer of clay – fragments of quartz and ash from an explosive volcanic eruption. On top of this are dark mudstones, rich in organic matter, a telltale sign of low-oxygen conditions on the sea floor. There is also iron pyrite (fool's gold), again indicating sulphurous, low-oxygen conditions. The former abundance of fossils has gone. Where once hundreds of species lived in a complex web, now only a few isolated shells survive, wedged in the mud. Most life in the sea has been wiped out. And according to geologists working on the Meishan section, the whole catastrophic event is recorded within a mere 12 mm of strata.

More secrets lurk in the Meishan rock beds. The band of volcanic ash enabled precise dating based on the degradation of uranium isotopes into lead – indicating that the event happened 251 million years ago. Carbon isotopes also change, showing that something had gone seriously awry with the biosphere and the carbon cycle. A clue as to why comes from the oxygen isotopes, which also show a dramatic shift between oxygen-16 and oxygen-18 – indicating a big fluctuation in temperatures. Here, perhaps, is the most explosive revelation of all: the temperature had risen, not by one, two or even four degrees. It had shot up by no less than six degrees. The end-Permian extinction, it seems, took place at a time of rapid greenhouse warming.

Outside China, other exposures of Permian–Triassic boundary rocks tell a similar story of global apocalypse and doom. In northern Italy, marine sedimentary layers from an end-Permian coastal area contain soil materials washed off the land in a catastrophic bout of soil erosion. Under normal circumstances plants anchor the soil, protecting it from being eroded away by the rain. But this failed to happen, and the conclusion is stark: almost the entire plant cover of the land must have been removed. Something had scythed through forests, swamps and savannahs, and once the monsoon rains arrived, there was nothing left to hold on to the precious soil, which washed in great torrents into the ancient ocean.

Where dead vegetation was left on the land, it simply rotted in situ. A 'fungal spike' has been reported from the rocks of Israel's Negev Desert and other places around the world, the preserved spores from a proliferation of toadstools which sprouted quickly on the dying trees and shrubs. For organisms which feast on the dead, it was a time of abundance and plenty.

In modern-day South Africa's Karoo Basin, researchers hunting for fossils spanning the Permian–Triassic boundary came across an unusual layer from the time of the extinction. This 'event bed', again indicating catastrophic erosion, is lifeless: it contains no fossils yet discovered. It also indicates a dramatic change in the climate, from wet to arid, at just the time all the life forms disappeared. What was once a deeply carved river valley, with abundant life on its banks, became a braided series of channels in a drought-stricken landscape. With no vegetation to hold its banks together, the river meandered across the emerging desert. What had once been a Garden of Eden had been transformed into Death Valley.

Further indications of an apocalyptic super-greenhouse come from the unlikely location of Antarctica. Nearly three thousand metres up on Graphite Peak in the central Transantarctic Mountains, preserved soils show sharply increased rates of chemical weathering, most likely because higher levels of CO_2 in the atmosphere were making rainfall more acidic. Importantly, the rock sediments also show that the transition from normal into greenhouse conditions was – by geological standards at least – spectacularly rapid, in the order of 10,000 years or less.

Geologists David Kidder and Thomas Worsley propose a fascinating model for how this greenhouse world worked, and also how it came about. The seeds were sown, they suggest, tens of millions of years before the extinction event, when tectonic mountain building ceased and the lack of chemical weathering allowed atmospheric carbon dioxide to build up gradually to dangerous levels. As the Permian drew to a close, CO_2 concentrations were four times higher than today's, giving a big boost to global temperatures.

Like a deadly game of dominoes, the climatic transition brought with it a chain of positive feedbacks, each of which added to the crisis. Desert belts expanded, whilst forests retreated to cooler refuges close to the poles, further reducing drawdown of CO_2 by photosynthesis. Deserts reached 45 degrees north (central Europe and northern USA in today's continental configuration), and perhaps invaded even as far north as 60 degrees, close to the Arctic Circle. These deserts would have been unimaginably hot, and high levels of evaporation from coastal oceans would have left seawater salty and dense, dragging warm water down to the ocean depths. This is the reverse of what happens in today's world, where it is cold water at the poles that sinks into the ocean abyss. But with hotter poles in the Permian greenhouse, sinking in the polar oceans slowed down and ceased.

Warm water may be nice to swim in, but as we saw earlier, once widespread across the oceans it is a killer. Less oxygen can dissolve in warmer seas, so throughout the water column conditions became gradually stagnant and anoxic. Oxygen-breathing water-dwellers – all the higher forms of life from plankton to sharks – faced suffocation. Warm water also expands, and sea levels rose by 20 metres during the Permian crisis, flooding continental shelves and creating shallow, hot seas as the anoxic waters swept outwards onto the land surface.

These hotter oceans would have spawned hurricanes of staggering ferocity, far outdoing anything we see today. Modern storms are limited by cold water both at depth and at higher latitudes, but in the end-Permian greenhouse warm oceans spanned the world from pole to pole. Super-hurricanes (sometimes termed 'hyper-canes') would have had enough fuel to carry them to the North Pole and back, perhaps even allowing them to repeatedly circum-navigate the globe. Only dry land would have stopped them, but a hypercane hitting a coast would have triggered flash floods that no living thing could have survived. These stupendous storms would also have delivered vast amounts of heat to high latitudes, increas-ing the greenhouse effect still further through water vapour and cloud feedbacks.

But this was still only the beginning. A balanced Earth system might have withstood such shocks. Unfortunately, however, fate had decided differently. Just as the Permian greenhouse intensi-fied, a giant plume of magma was making its way upwards towards the Earth's crust from the molten mantle, aimed like a knife at the heart of Siberia. When it reached the surface, molten rock burst forth with spectacular violence, ejecting ash and volcanic debris for hundreds of miles, and blotting out the sun with dust and sul-phur dioxide. As more and more magma erupted over the millen-nia, it built up in layers many hundreds of metres thick, over an area larger than western Europe. With each successive eruption, more basalt flooded over the land, releasing billions of tonnes of CO_2 from hellish gaping fissures in the Earth's surface.

Life might have better survived the Siberian flood basalts had they not come at a time when intense warming was already push-ing the biosphere towards the margins of survivability. As it was, the eruptions were a further body blow, releasing poisonous gases and CO_2 in equal measure, sparking torrential storms of acid rain at the same time as boosting the greenhouse into an even more extreme state. As Michael Benton outlines in his book *When Life Nearly Died*, these monsoons of sulphuric acid would have further stripped the land of vegetation, washing rotting tree trunks and dead leaves into the already stagnant oceans. By this time much of life would have been dead or dying. Creatures in deep burrows might have survived the initial worst of the crisis, but anything emerging onto land would soon die from heat or be laid low by starvation. With most vegetation – the base of the food chain both on land and in the oceans – wiped out, little else could have survived for long. Oxygen levels in the atmosphere plunged to 15 per cent (contemporary levels are 21 per cent) – low enough to leave any fast-moving animal gasping for breath even at sea level.

Worse was still to come. With warm water rapidly reaching the ocean depths, a now-familiar monster was stirring on the continental shelves – methane hydrates. Runaway global warming had begun.

Much of the initial methane release would have dissolved in the water column, accumulating gradually over time. As the stream of bubbles kept rising, however, each successive layer of water would have gradually reached saturation point. Once this explosive was primed, all that would be needed was a detonator.

This is how events unfold: First, a small disturbance on the sea floor drives the methane-gas-saturated parcel of water upwards. As it rises, bubbles begin to appear, as dissolved gas fizzes out with the reducing hydrostatic pressure – just as a bottle of fizzy lemonade overflows if the top is taken off too quickly. These bubbles make the parcel of water still more buoyant, accelerating its rise through the water column. As the water surges upwards, reaching explosive force, it drags surrounding water up with it, spreading the process. At the sea surface, water is shot hundreds of metres into the air as the released gas blasts into the atmosphere. Shockwaves propagate in all directions, triggering follow-up eruptions nearby.

None of this is theoretical conjecture – a similar process in miniature took place as recently as 1986 in Cameroon's Lake Nyos, where volcanic outgassing under the lake bed is continually releasing carbon dioxide. On the late evening of 12 August 1986, this gas erupted, creating a gas–water fountain up to 120 metres in height and releasing a lethal carbon dioxide cloud which asphyxiated 1,700 people in the surrounding area. Being heavier than air, the CO_2 cloud hugged the ground, suffocating many of the victims in their beds.

A methane cloud might behave in very much the same way. Loaded with water droplets, it spreads out over the land surface like a toxic blanket. Moreover, unlike CO_2, methane is flammable. Even in air–methane concentrations as low as 5 per cent, the mixture could ignite from lightning or some other source of sparks, sending fearsome fireballs tearing across the sky. A modern comparison might be the fuel–air explosives used by the US and Russian armies, whose destructive power is comparable with tactical nuclear weapons. These so-called 'vacuum bombs' spray a cloud of fuel droplets above a target (ideally an enclosed space like a cave) and then ignite it, sucking out the air and sending a blast wave strong enough to kill and injure over large areas.

The methane–air clouds produced by oceanic eruptions would dwarf even the most severe modern battlefield weapons, however, and explosions in the largest clouds could generate explosive blast waves able to travel faster than the speed of sound. With a supersonic blast, it is the pressure from the shockwave itself which ignites the mixture, pushing out an explosive front at speeds of 2 kilometres per second and vaporising everything in its path.

The likely effects on the animals and plants that inhabited the Permian world are scarcely imaginable. A comparatively small eruption of oceanic methane could therefore quickly become a very effective agent of mass extinction. As the chemical engineer Gregory Ryskin writes, in a paper specifically addressing 'kill mechanisms' at the end-Permian, this methane 'could destroy terrestrial life almost entirely'. A major oceanic methane eruption, he estimates, 'would liberate energy equivalent to 10^8 megatonnes of TNT, around 10,000 times greater than the world's stockpile of nuclear weapons'. This global conflagration might even cause short-term cooling akin to a nuclear winter, before boosting global warming further with the CO_2 produced by the combusted methane. (And any uncombusted methane would have an even more serious warming effect.)

The methane killing agent may not have acted alone. As vegetation and animal carcasses rotted in the stagnant oceans, large quantities of hydrogen sulphide were building up in the depths. Evidence of this sulphurous ocean is still preserved in Permian rocks in eastern Greenland, where telltale pyrites are common amongst the black shales laid down at the time of the catastrophe. Poisonous at minute concentrations (and smelling of rotten eggs), any release of hydrogen sulphide into the atmosphere would have laid low animals left alive in any places which somehow escaped the methane eruptions.

Just as importantly, the sulphurous brew in the oceans would also have been a very effective agent of extinction in the marine realm, killing all oxygen-breathing life forms. As if this somehow weren't enough, the hydrogen sulphide cloud would also have attacked and destroyed the ozone layer, letting in dangerous ultraviolet radiation from the Sun. Disfigured spores have recently been discovered in the same Permian rocks of eastern Greenland, indicating that surviving land plants may indeed have been suffering from DNA mutations caused by prolonged UV exposure.

At very high concentrations, methane also destroys ozone: one modelling study specifically investigating conditions at the end-Permian found that if surface methane concentrations reached 5,000 times background levels – such as might be the case during a large-scale methane hydrates eruption – half the ozone column would be destroyed, increasing UV radiation to the surface by a factor of seven. This in itself could be a major cause of the extinction, the authors suggest. Moreover, the two agents combined – hydrogen sulphide and methane – could massively increase the destructive effect on the ozone layer.

With all these successive disasters let loose upon the Earth, it is hardly surprising that the end-Permian mass extinction outranks all others. According to some calculations, 95 per cent of species – on land and in the sea – were wiped out. In the oceans a few shells stuck it out, wedged deeply in the mud. On land, only one large vertebrate made it through the extinction bottleneck: the piglike *Lystrosaurus*, which for millions of years afterwards had the planet pretty much to itself. There is a distinct 'coal gap' in the early to mid-Triassic, the geological period following the Permian, showing that only sparse vegetation survived, with nothing like the lush forests that laid down the thick coal seams of the earlier Permian and Carboniferous periods. It took 50 million years – well into the Jurassic – before anything like pre-extinction levels of biodiversity returned.

Back to the future

All geologists agree that the end-Permian crisis was the mother of all disasters. So what lessons might it have for us if our world heads towards six degrees of warming? Clearly, given the 251-million-year time gap that has since passed, events cannot be expected simply to repeat themselves. The continents are arranged differently for a start, perhaps allowing for a better circulation of the oceans. There is more oxygen in the atmosphere, so even the worst rates of greenhouse warming are unlikely to asphyxiate us. Nor am I suggesting that the prospect of methane fireballs tearing across the sky is remotely likely in today's world – it is merely one of many theories that have been advanced to explain the end-Permian die-off, and as such may hint at the worst the greenhouse Earth can sometimes do to its inhabitants.

On the other hand, there are some aspects of today's global warming crisis which are especially worrying, even when contrasted with the horrors that struck the world at the end-Permian. An extinction event – tentatively titled the Anthropocene Mass Extinction – is already under way, largely independently of global warming. With so many plants and animals drastically reduced and forced onto the margins of survival, the natural world is already less resilient to change than would have been the case in the late Permian. Consider our cousins the great apes, so small in number now that more human babies are born every day than the entire global population of gorillas, chimpanzees and orang-utans put together. With so many species of life already hanging on by their fingernails thanks to *Homo sapiens*' rapacious ways, just moderate climatic changes will be enough to push them over the cliff.

Human disturbance also makes it much harder for animals and plants to migrate and adapt, as we saw in previous chapters. What remains of nature is under siege in island 'reserves' between agricultural and urban deserts. As the temperatures rise and real deserts shift into the mid-latitudes, these islands of nature will be submerged one by one, snuffed out by the changing climate.

Consider also the rate of change. Even the fastest rates of volcanic outgassing of CO_2 take millennia to have any measurable effect on the climate. We're accomplishing the same feat in decades. As stated earlier, the end-Permian greenhouse probably took at least 10,000 years to play out. We could achieve the same level of warming in a century, a hundred times quicker even than during the worst catastrophe the world has ever known. Even given the uncertainties of the geological record, it is difficult to state this point strongly enough: human releases of carbon dioxide are possibly happening faster than any natural carbon releases since the beginning of life on Earth. The Palaeocene–Eocene Thermal Maximum and the Cretaceous 'ocean anoxic events' all apparently saw greenhouse gases rise more slowly than is currently the case. In sheer volume terms of carbon, we are still far below their magnitude, but the speed of change is already unprecedented, and puts us into uncharted territory.

Of course, our carbon releases are not malicious – for most of us, our huge energy consumption is simply a part of modern life – but to the biosphere this hardly matters. If we had *wanted* to destroy as much of life on Earth as possible, there would have been no better way of doing it than to dig up and burn as much fossil hydrocarbon as we possibly could.

Many people instinctively feel that small creatures like we humans cannot really have any serious impact on a very big object like the planet. But if you doubt the scale of the enterprise that human society is currently involved in, go and stand by the side of a busy motorway, and then look up at the sky. Remember that the breathable atmosphere extends a mere 7,000 metres above your head. Then think of how many other motorways there now are criss-crossing the globe, from Bangkok to Berlin, each chock-full of cars and trucks, with each vehicle's exhaust pipe continually exhaling its deadly brew of carbon dioxide and other gases. Remember too to add in all the power stations, the aircraft, the home boilers and the gas fires, and remember that this situation goes on day and night, 24/7 across the whole of the globe.

Better still, look at a satellite composite photo of the Earth at night, see each continent lit up by a tangled spider's web of cities, and marvel at the visual totality of this plethora of continual human energy consumption, 80 per cent of which is based on burning fossil fuels. Then it might seem less surprising that each year's CO_2 concentrations are higher than the last, and that each breath you inhale has more carbon dioxide in it than any breath ever taken by any human before you over the entire evolutionary history of our species. It can hardly be a surprise either that the climate is changing rapidly: what *would* be a surprise were if everything continued as normal.

So the lesson of the end-Permian is this: the planet can rapidly turn very unfriendly indeed once it is pushed far enough out of kilter. Today, vast volumes of methane hydrates are again lodged on subsea continental shelves, biding their time for the trigger of rising ocean temperatures. Just how far they can be safely pushed, no one can tell.

Nor is there any reason to rule out ocean stratification and hydrogen sulphide poisoning as another possible disaster scenario. The gradual shutting down of the Gulf Stream could be just one cog in this much larger machine: as the oceans stopped circulating, warmer waters would penetrate into the depths, carrying less dissolved oxygen and gradually eliminating aerobic life. Stagnant oceans would be largely invisible to us land dwellers: indeed, much of the Black Sea today is anoxic, but cold oxygen-rich surface waters keep a lid on the poisonous liquid beneath.

However, catastrophic hydrogen sulphide release from the ocean depths is more than just a speculative possibility – it happens occasionally in today's world, albeit on a much smaller scale, off the coast of Namibia. Here poisonous sulphur erupts upwards from rotting organic material on the sea floor in massive bursts, the resulting toxic brew discolouring the sea surface over such large areas that the phenomenon is easily observed from satellites. Huge numbers of fish – indeed everything living in the oceans nearby – dies as a result, and on the nearby shorelines human inhabitants suffer with the unpleasant smell and corrosive effect of toxic hydrogen sulphide gas. According to scientific investigations, the eruptions are mostly driven by sea-floor methane – which effervesces up through the water column into huge events exactly as is suggested might have happened on a worldwide scale at the end of the Permian.

Indeed, Andrew Bakun and Scarla Weeks, both experts on the Namibian hydrogen sulphide phenomenon, suggest specifically that global warming might prompt a wider-scale release of the poisonous gas by increasing oceanic upwelling. Vulnerable locations, where deep water already comes to the surface – though thankfully at present without toxic gas eruptions – include the coasts of Morocco, Mauritania, Peru and California. Were this unwelcome change ever to begin, however, coastal dwellers would be the first to sniff the rotten eggs: the human sense of smell can pick up hydrogen sulphide in the parts per trillion range. (At higher concentrations the olfactory nerve becomes paralysed and we lose the ability to detect the toxin.) It would be a silent killer: imagine the scene at Bhopal following the Union Carbide gas release in 1984, replayed first at coastal settlements, then continental interiors across the world. At the same time, as the ozone layer came under assault, we would feel the Sun's rays burning into our skin, and the first cell mutations would be triggering outbreaks of cancer amongst anyone who survived.

Could humanity itself ever go extinct? I think it unlikely: people in general have a unique combination of intelligence and a strong survival instinct. Humans will go to extraordinary lengths not to die, as countless true stories of survival against the odds attest. I myself have crawled down an Andean mountain in a state of delirious semi-consciousness when the easiest thing by far would have been to lie back and let go – but the survival instinct was of course too strong. Even given the most dramatic rates of warming imaginable, somewhere, surely, it will still be possible to raise crops and grow food. Rainfall will not stop, and the melting ice sheets will provide plentiful supplies of water in polar regions. Feeding a world of eight or nine billion is another matter, but the idea that every single one of us could be wiped out strikes me as inconceivable. Unlike the Permian terrestrial animals, we can stockpile enough food in preserved form to last for many years. We can create artificial atmospheres to isolate ourselves from what is going on outside. As some scientists have suggested, we could take emergency measures to geo-engineer our climate, perhaps by spreading solar mirrors in space or scattering sulphates through-out the upper atmosphere in a last-ditch attempt to cool things down. We could one day even set up colonies on other planets.

And yet somehow this is scarce consolation given the torments that may lie in store. Extreme global warming may not be a sur-vival crisis for humanity as a species, but it will certainly be a survival crisis for most humans unfortunate enough to inhabit a rapidly warming planet, and that is surely bad enough. Stalin was wrong to say that a million deaths is merely a statistic: that is why he is still hated today. Every human death, of every baby, every mother, every brother, father or sister, will be a unique tragedy for which the whole world should grieve: not least because such an outcome is still today avoidable.

Many people, when confronted by such awful possibilities, take refuge in a sort of geological fatalism: the oft-heard refrain that life will go on, with or without us, and that at the end of the day it doesn't really matter. The planet might even be better off without *Homo sapiens*, some might suggest. Notwithstanding the moral questions that this sort of attitude raises (it is a bit like saying the Nazi Holocaust didn't matter because the high post-war birth rate soon replaced the six million dead), it is far from clear that life *will* always go on. The Sun is getting hotter as it burns up its finite supply of nuclear fuel, and for millions of years into the future our planet's great challenge will be to keep itself cool as the output of solar radiation inevitably increases.

This is a dangerous time to be fiddling with the Earth's thermostat. Scientists have calculated that only a billion years remain before the biosphere will go extinct for ever from overheating – the planet is already 4.6 billion years old, and for much of this time it was lifeless. The tens of millions of years it will take for new forms of life to re-establish, and for biodiversity to re-evolve into new complex ecosystems after any human-caused mass extinction, is a sizeable portion of this remaining habitable time. As James Lovelock writes, 'Mother Earth' is now an old lady in her sixties, no longer as resilient as she once was. With our conscious actions, we are now measurably shortening her lifespan.

So far as we yet know, this is the only planet in the entire universe which has summoned forth life in all its brilliance and variety. To knowingly cut this flowering short is undoubtedly a crime, one more unspeakable even than the cruellest genocide or most destructive war. If each person is uniquely valuable, each species is surely more so. I can see no excuses for collaborating in such a crime. As the post-war Nuremberg trials established, ignorance is no defence; nor is merely following orders. To me the moral path lies not in passively accepting our destructive role, but in actively resisting such a horrendous fate.

As I stated at the beginning of this book: nothing in the future is set in stone; we still have the power – though it diminishes every day – to alter the ending of this terrible drama. It need not yet end in tragedy, and to make this point more clearly the next and final chapter will examine our options for avoiding each successive degree rise in temperature. There, and only there, hope lies.

As Dylan Thomas wrote:

Do not go gentle into that good night,
Old age should burn and rave at close of day;
Rage, rage against the dying of the light.

7

7

CHOOSING OUR FUTURE

> 'We see, like those with faulty vision,
> things at a distance,' he replied. 'That much,
> for us, the mighty Ruler's light still shines.
> When things draw near or happen now,
> our minds are useless. Without the words of others
> we can know nothing of your human state.
> Thus it follows that all our knowledge
> will perish at the very moment
> the portals of the future close.'
>
> Dante, *Inferno*, Canto X: the Sixth Circle of Hell

Like Farinata, the ghostly apparition who speaks to Dante the resonant passage above, it is much easier for all of us to see events when they still lie at a distance. It is also much easier for us to affect them – for unless we decide to reduce greenhouse gas emissions within just a few years from now, our destinies will be chosen and our path towards hell perhaps unalterable as the carbon cycle feedbacks detailed in earlier chapters threaten to kick in one after another. Like the tormented souls Dante meets in the

Sixth Circle of Hell, once the 'portals of the future close' – in Amazonia, Siberia or the Arctic – we may find ourselves powerless to affect the outcome of this dreadful tale.

The warning is clear – but do we have the collective will to hear it? In November 2006, scientists working on the Global Carbon Project announced that emissions were then rising four times faster than they were a decade ago. In other words, all of our efforts – of carbon trading, switching off lights, the Kyoto Protocol, and so on – have had a discernible effect so far: *less than zero*. The project participants point out that we are getting further and further away from any of the IPCC's 'stabilisation pathways' with each day that passes. Things do not look good.

Moreover, due to the thermal time lag of the planet, even if atmospheric concentrations of greenhouse gases quickly stabilise (which – let's be honest – they aren't going to), temperatures will rise by between 0.5 and 1 degree Celsius, whether we repent or not. Like the walking dead that Dante meets in the First Circle of Hell, the Alpine glaciers, the Nebraskan grazing lands and the resplendent coral reefs are already condemned by events which lie in the past. The latest studies also make it clear that we are very close to a tipping point in the Arctic, and that only a massive international effort can now save the Arctic ice cap from complete collapse – leaving an ice-free summer North Pole as early as 2040. Indeed, if we are to believe the evidence of the Pliocene, then as much as three degrees of warming, with eventual sea level rises of 25 metres, is on the cards even at current concentrations of CO_2. That would take centuries to play out, however, making this a slightly less pessimistic analogy than it may first appear.

According to sophisticated computer modelling of future rates of climate change, we do have a short time left to cut back emissions in order to avoid 'dangerous' levels of warming, and can still aim for a 'safe landing' within the 1–2°C corridor. This window of opportunity is very nearly closed, however. My conclusion in this book, one which is supported by the 2007 IPCC report, is that we have less than a decade remaining to peak and begin cutting global emissions. This is an urgent timetable, but not an impossible one. It seems to me that the dire situation that we find ourselves in argues not for fatalism, but for radicalism.

Knowing what we don't know

To borrow from Donald Rumsfeld, one of the great 'known unknowns' in the climate change field, and the only part of the climatic equation that we actually have any control over, is future emissions – just how many more billions of tonnes of greenhouse gases humans are going to emit through burning fossil fuels and clearing forests over years to come. The IPCC has developed a complex set of 'emissions scenarios', each one of which builds in different assumptions about economic and population growth, globalism versus localism, and other key factors over the next century. They are 'scenarios' rather than 'predictions' because we simply have no idea which is most likely to come true, and the IPCC explicitly assigns no probabilities in its analysis: each scenario is considered equally likely.

Emissions scenarios are actually a much bigger unknown than any of the much-discussed climate science uncertainties which contrarians are always harping on about. These uncertainties at least refer to physical realities that scientists should eventually be able to narrow down given enough brain and computer power. Future emissions, however, depend on decisions that are yet to be taken billions of times over by the likes of you and me. They depend on economics and politics, rather than the more solid ground of physics. For this reason, long-term emissions can probably never be accurately predicted.

The second 'known unknown' is a genuinely scientific one: what academics call 'climate sensitivity'. (I should note that neither of these uncertainties affects the scenarios portrayed in previous chapters of this book, which refer to the impacts of actual temperature rises, independently of the emissions scenarios that generate them. I am raising them here in order to begin tackling the crucial question which this chapter sets out to answer: which emissions paths will lead to which temperature outcomes?) Climate sensitivity is defined in the technical parlance as the equilibrium temperature response of the planetary system to a doubling of pre-industrial atmospheric concentrations of CO_2. It is important because if climate sensitivity is low (i.e. the climate is not very sensitive to carbon), then high carbon emissions will lead to relatively manageable temperature rises. On the other hand, if climate sensitivity is high, then even drastically curtailing emissions may not save us from rapid increases in temperature.

Climate models, because they are designed by different teams using slightly different parameters, come up with slightly varying figures for this crucial number. This is a lot more than just guesswork: models, after all, are based on physical laws (although many of these are imperfectly understood), and most of them converge at around 3°C – implying that a doubling of pre-industrial carbon dioxide levels, from about 280 to 550 parts per million, would yield an eventual temperature rise of three degrees.

More recently, however, some scientific teams have suggested that the true value for climate sensitivity could be much higher. One of the best-known is the team behind Oxford University's climateprediction.net project, which involved thousands of people downloading and running a climate model on their home computers. Each downloaded model had tiny differences from the next, and by examining thousands of results the researchers hoped to get a cluster of 'best estimates' for climate sensitivity. They made international headlines in the process: while most of the home-run model results came in at the standard 3°C or so, a fair few showed climate sensitivities as high as an incredible 11°C. As the project coordinator Dave Frame told reporters: 'The possibility of such high responses has profound implications. If the real world response were anywhere near the upper end of our range, even today's levels of greenhouse gases could already be dangerously high.'

One of the biggest hurdles to calculating climate sensitivity is the phenomenon of 'global dimming'. It has long been known that sulphate aerosols – released during the combustion of fossil fuels, and an important contributor to acid rain – have a short-term cooling effect on the climate, temporarily shielding us from the full effects of greenhouse warming by cutting off some of the Sun's rays. These aerosols are the most likely culprits for the perplexing small drop in twentieth-century global temperatures between 1940 and about 1960, at a time when greenhouse gas emissions were beginning to rise rapidly.

Most projections suggest that global dimming will decline over future decades as pollution-control measures take effect – indeed, this process is probably already under way, perhaps explaining the more rapid global warming measured since the 1980s. Aerosols don't stay very long in the atmosphere: they are scrubbed out by rainfall in a matter of days, whereas CO_2 lasts on average for a century. Once this aerosol shield is removed, temperatures may well then follow the upper extremes of the IPCC's projections – closer to six degrees than to two, according to the latest studies.

Moreover, the gap between models and palaeoclimate remains troubling. As mentioned earlier, CO_2 levels similar to today's seem to have caused around 3°C of warming during the Pliocene. Likewise, current models cannot reproduce the Eocene hothouse event, when the Arctic Ocean was as balmy as the Mediterranean, without unrealistically high levels of simulated CO_2. The same problem repeats itself during the Cretaceous greenhouse. If estimated Cretaceous CO_2 levels are inputted into climate models, the levels of simulated warming are much less than have been estimated by geologists as having really taken place 100 million years ago. So either the fossil data about past temperatures and CO_2 levels is wrong, or the models are underestimating climate feedbacks. They can't both be right.

Climate models, by their very nature, are part of a reductionist scientific enterprise, picking apart the different bits of the climate system and trying to understand and describe them individually using physical equations. Put all these equations together, the logic goes, representing everything from clouds to sea ice, and you get your model. But the whole of a complex and interacting system is not necessarily just a sum of all the parts. In a crude comparison, that is why doctors cannot reassemble a fully functioning human body by sewing together skin, ears, teeth, blood, nerves and bones from a dissected human donor. The secret lies in the way all human chemical and biological components interact within a living organism. The same is true for the planetary 'organism', which – as Gaia theory shows – exhibits many of the same self-regulatory characteristics as a living body.

The missing link could be climate feedbacks and interactions which aren't in the models. Whilst some better-understood elements of the carbon cycle are now included, there are many feedback uncertainties which simply cannot be quantified. No one knows, for example, precisely how much CO_2 and methane could be released by melting permafrost. Nor does anyone know for sure when the methane hydrates trigger may be pulled, nor how these processes might have occurred during the Cretaceous. Jim Hansen may have partly resolved this conundrum by distinguishing between the timescales involved in the Earth's feedback processes. 'Fast feedbacks', he points out, include changes in water vapour, clouds, atmospheric dust, sea ice and snow cover, and these are well accounted for in the models. Add them all together, and you get the generally agreed figure of about 3°C for a doubling of CO_2. But slower feedbacks – such as ice sheet collapse, and greenhouse gas changes resulting from alterations in vegetation and the carbon cycle – are not well modelled, and increase climate sensitivity substantially over longer timescales.

The modellers are beginning to catch up, however, and these different feedback effects are now being studied in detail. One May 2007 study reveals, as expected, that models which include carbon cycle feedbacks tend to dramatically increase their projected rates of warming. Another looked at how greenhouse gas feedbacks work during the transition into and out of ice ages. Using this real-world data, they projected that the IPCC's 2001 upper value for twenty-first-century warming, 5.8°C, 'could be increased to 7.7°C, or nearly 2°C of additional warming'. A second international study, using temperature information from the last thousand years, reaches essentially the same conclusion. Both suggest that the discrepancy between palaeoclimate and models will eventually be resolved in favour of palaeoclimate – not a reassuring prospect. Six Degrees? Maybe make that Eight Degrees.

Setting a target

As we saw earlier, it is very likely that between 0.5 and 1 degree of further warming is already in the pipeline due to the thermal time lag of the Earth system, so even stopping a rise in CO_2 tomorrow would still see us enter the 'one-degree world'. Temperatures will rise for another thirty years from now even if greenhouse gas emission cuts are implemented instantly – because of the massive amount of carbon we have already dumped into the atmosphere.

But if we still have time to stabilise the climate within two degrees – and models suggest we do – then we can potentially save great swathes of global biodiversity, slow the melting of Greenland and associated sea level rise down to tolerable levels, and avoid the most dangerous positive feedbacks that could kick in as we approach three degrees of warming.

It is this latter objective which is so utterly vital, and which is perhaps the central message of this book. If, as chapter 3 showed, we cross the 'tipping point' of Amazonian collapse and soil carbon release which lies somewhere above two degrees, then another 250 parts per million of CO_2 could pour into the atmosphere, yielding another 1.5°C of warming and taking us straight into the four-degree world. Once we arrive there, the accelerated release of carbon and methane from thawing Siberian permafrost will add even more greenhouse gas to the atmosphere, driving yet more warming, and perhaps pushing us on into the five-degree world. At this level of warming, as chapter 5 showed, oceanic methane hydrate release becomes a serious possibility, catapulting us into the ultimate mass extinction apocalypse of six degrees.

The lesson is as clear as it is daunting: if we are to be confident about saving humanity and the planet from what could be the worst mass extinction of all time, worse even than that at the end of the Permian, *we must stop at two degrees.*

So can we make it? One academic analyst has calculated that there is only a 7 per cent chance that we already have crossed the two-degree line. This is already a worryingly high figure: I would personally not set foot on a ship that had a 7 per cent chance of sinking in mid-ocean, for example. Still, marooned as we are on the only life-supporting planet we know of in the universe, we don't have a lot of choice.

However, to look on the bright side, if we accept this figure then the world still has a 93 per cent chance of coming in under two degrees – but only at current atmospheric carbon dioxide concentrations. Every year that we allow CO_2 levels to go on rising, the odds of overshooting two degrees go on shortening. Within seven years, as CO_2 concentrations climb to 400 ppm (they stood at 382 ppm in 2007, and rise at about 2 ppm per year), our room for manoeuvre will have narrowed considerably. Even if we assume a relatively low climate sensitivity, we will by then have a less reassuring 75 per cent chance of meeting the two-degree goal. These are not odds that many people would accept for anything other than a very low-stakes game of poker – they are certainly not good odds when one is gambling the planet.

Still, I repeat: we don't have much choice. It is hardly realistic to demand that greenhouse gas emissions come down by 60 per cent or so within the next decade, as would be required to stabilise atmospheric CO_2 concentrations below 400 ppm. Stabilise at 550 ppm, as the British economist Sir Nicholas Stern suggests is the only politically realistic option, and our chances of staying below two degrees are slim – less than 20 per cent. With CO_2 levels that high, avoiding even three degrees becomes increasingly tricky. Indeed, according to the academics, stabilising CO_2 levels at 550 ppm leaves a 10 per cent chance of overshooting even four degrees, whilst the odds of staying below three degrees at that CO_2 concentration are no better than fifty-fifty. (Don't worry: I'll summarise all this in a table below.) Clearly, the target we need to aim for depends on what level of risk we are prepared to accept – bearing in mind that failure could mean runaway global warming and the destruction of most of life on Earth.

Most of this discussion will sound complex and arcane to just about everyone except specialists in the subject. But it shouldn't. This is actually the key question currently facing humanity – far more important than terrorism, crime, healthcare, education or any of the other everyday concerns that fill up our newspapers and television screens. We all face a decision: what temperature, and by implication, what CO_2 concentration, do we aim for? 400 ppm? 550 ppm? My guess is that a standard opinion poll of the general public would not pick up too many answers. Most of the population would be in the 'don't know' category. Even the political class is only just beginning to understand the relevant terminology, and no major political party anywhere in the world that I know of has a definite policy on this matter. Ultimately, however, this is a political decision, and one in which all the world's people must be able to participate in an informed, democratic way if the decision is to be observed and supported by everyone.

This issue also raises a more profound question about our place on the Earth. We humans, one species of animal amongst millions, have now become de facto guardians of the planet's climate stability – a service which used to be provided free (given a few ups and downs) by nature. Without realising it, we have appointed ourselves janitors, our sweaty ape hands resting heavily on the climatic thermostat. A more awesome responsibility can scarcely be imagined.

So what are the specifics of a two-degree target? As I mentioned earlier, in order to be 75 per cent certain of temperatures staying below the magic two-degree threshold, global greenhouse gas emissions need to peak within the next *seven years*, by 2015. Emissions must then continue to decline, and by 2050 they must have fallen by 85 per cent. This will stabilise CO_2 concentrations at around 400 parts per million (or 450 ppm 'CO_2-equivalent' – that is, if the warming effect of all greenhouse gases like methane and nitrous oxide is expressed in CO_2 terms). If concentrations rise above this level, the chances of coming in under two degrees diminish accordingly. This is a conclusion I base on my own reading of the latest science, but which is supported by the 2007 IPCC report – which also makes clear that global emissions must peak by 2015 if CO_2 concentrations are to remain at or below 400 ppm, and the associated temperature rise limited to two degrees.

Here are my best estimates in tabular form, to try and simplify the above discussion. In order to avoid each successive degree of warming with a high degree of likelihood (75 per cent), I estimate that we need to take the following action:

Degree change	Actual temperature in Celsius	Action needed	CO_2 target
One degree	0.1–1.0°C	Avoidance probably not possible	350 ppm (today's level is 380 ppm)
Two degrees	1.1–2.0°C	Peak global emissions by 2015	400 ppm
Threshold for carbon cycle feedback?			
Three degrees	2.1–3.0°C	Peak global emissions by 2030	450 ppm
Threshold for Siberian methane feedback?			
Four degrees	3.1–4.0°C	Peak global emissions by 2050	550 ppm
Five degrees	4.1–5.0°C	Allow constantly rising emissions	650 ppm
Six degrees	5.1–5.8°C	Allow very high emissions	800 ppm

The table illustrates how hopelessly inconsistent current climate policies are – even of some major environmental groups. The European Union has mentioned (though not formally adopted) a 550 ppm target, whilst at the same time demanding that global temperature rise be kept below two degrees. In all likelihood, as the table shows, 550 ppm means four degrees plus additional positive feedbacks. Friends of the Earth, a well-respected and dedicated environmental group which has done brilliant work to raise awareness of climate change, also wants to see global temperatures stay this side of two degrees – for the same reasons as I do. But its political campaigning only demands a CO_2 stabilisation of 450 ppm, despite the fact that this carbon concentration, according to the best available science, gives a 75 per cent chance of *missing* its two-degree target. Many other groups are caught on the horns of the same dilemma: that only by advocating 'politically unrealistic' CO_2 concentrations can extreme global warming be reliably avoided. But then what is politically realistic for humans is wholly unrelated to what is physically realistic for the planet.

So if our target is two degrees, in order to confidently avoid the unstoppable climatic domino effect of positive feedbacks, global emissions of all greenhouse gases must peak by 2015, and drop steadily thereafter with an ultimate CO_2 stabilisation target of 400 ppm (or 450 ppm for CO_2-equivalent) – however politically unrealistic this emissions trajectory might seem. The actual percentage emission cuts this target implies will depend on how the Earth's carbon cycle behaves, but in my view current science suggests that it means a worldwide 60 per cent cut by 2030, and an 85 per cent cut by 2050. This is again consistent with the projections from the IPCC's 2007 Fourth Assessment Report.

Given that per person carbon emissions vary hugely between different countries, no one should expect these percentage cuts to be adopted uniformly across the board. India, for example, currently emits about one tonne of CO_2 per person, China four tonnes per person, and the US twenty tonnes per person. Clearly if all these countries simultaneously adopted a 60 per cent emissions cut, basic inequalities would remain: India would emit 0.4 tonnes, China 1.6 tonnes, and the US eight tonnes. Any cementing of structural inequality such as this is obviously highly unlikely to form the basis of a successful global emissions agreement, for the simple reason that it would be grossly unfair.

So what would work as an across-the-board global agreement, given different levels of development? There is only one logical way out of the conundrum: rich countries must agree to trade their habituated inequity in return for poor countries' participation in an agreed climate regime, a compromise first proposed by the Global Commons Institute and known as 'Contraction and Convergence'. Under C&C, all countries would converge to equal per person emissions allocations by an agreed date, within the overall context of a contraction of global emissions to sustainable levels. It would be a historic bargain: the poor would get equality, whilst all (including the rich) would get survival.

In the US, where per person emissions are higher than in the majority of the world, this convergence process would require much heftier cuts than the global average, perhaps as high as 85 per cent by 2030, depending on the nature of the C&C deal. In order to make the system flexible and efficient, however, it is crucial that an international market in emissions permits is established – allowing poor countries to sell unused allocations to the rich, generating significant revenue in the process. This earning from a global carbon trade could help tackle poverty as well as ensuring that poorer countries have the option of pursuing a low-carbon development path.

So, to be blunt: the conclusion of this book is that we have only seven years left to peak global emissions before facing escalating dangers of runaway global warming. I am the first to admit that this target looks hopelessly unattainable. But it seems to me to be what the science demands, even based on fairly conservative assumptions – and assuming we are not over the line already. And let's be honest – reasons for wholesale pessimism are many and various: as I mentioned earlier, the rate of increase in global carbon emissions has quadrupled in the last decade. Across the world, emissions are still rising rapidly: the worst offender, the United States, now releases 16 per cent more CO_2 than in 1990, whilst China's total – though far lower in per capita terms – has accelerated by an even greater amount. According to some reports, China has already overtaken the US as the world's single biggest emitter.

Looking further forwards, the IEA projects that world energy demand will rise by more than half by 2030, with 80 per cent of the increase coming from fossil fuels rather than clean energy sources. Worryingly, this 'business as usual' scenario sees renewables increasing rapidly, but still accounting for only 2 per cent of total energy generation in 2030 because of the sheer scale of growth. Consequently, CO_2 emissions will rise by a frightening 52 per cent by the same date, according to this projection. Even the IEA now admits that these trends 'lead to a future that is not sustainable', and it therefore also advances a World Alternative Policy Scenario, 'in which energy-importing countries take determined action to cut demand and change the pattern of fuel use'. This optimistic scenario turns out on closer examination not to be so optimistic at all: CO_2 emissions are still projected to increase by a third by 2030, instead of the 60 per cent reduction we need if we are to meet the two degrees target.

It should now be clear that the business as usual trend (which we are currently on, irrespective of Kyoto) yields a high chance of reaching four, five and even six degrees of warming by 2100. Add in all of the uncertainties, and it should be abundantly clear that a further major growth in emissions takes the world into a gambling game in which the odds are stacked against us. And the higher emissions rise, the worse the odds get. Heads – global warming wins, tails – we lose. As one wag has pointed out, this is like playing Russian roulette with a Luger rather than a revolver. One bullet, one chamber – and we're pulling the trigger.

A reality check

Many books on global warming end with some rather platitudinous sentences about renewable energy, as if the authors believe – rather like Disney's Blue Fairy – that simply wishing for something and believing in it is enough to make it come true. My feeling is that if it were that easy to move away from fossil fuels, we would have done so already, or at least be heading rapidly in the right direction. Instead, as we saw above, the world is moving rapidly in the wrong direction, and seems set to continue doing so.

So should we despair about the prospects for reaching the two degrees target? No – but nor should we base policy on wishful thinking. We should perhaps start, therefore, by admitting openly that there is no quick fix. Fossil fuel energy has come to occupy a central role in almost every aspect of modern life, from transport to home heating, and accounts for 80 per cent of our energy supply. Indeed, our civilisation is defined by energy use more than by any other aspect of its nature: without massive inputs of energy, society would quickly grind to a halt, and billions would starve.

This was illustrated for me – in almost a farcical way – by the fuel protests which took place across the UK in autumn 2000, which led to petrol and diesel shortages for at most a couple of weeks, but in the process almost brought the country to its knees. Queues formed in supermarkets as consumers panic-bought groceries. People drove for miles to find petrol stations that hadn't run dry. In central Wales, cows stood unmilked in the fields at the same time as milk in plastic bottles sold out in all the shops. With the only usable churns now in the museum, even getting milk two hundred yards from farm to shop required big articulated lorries, which cart the product nearly a hundred miles to a centralised packaging and distribution point and then back again. And all that, of course, requires diesel.

This illustrates in minuscule how dependent we have become on fossil fuels, but the same problem can also be seen at the macro level. Jeffrey Dukes of the University of Utah is one of the few people to have done the maths, and his figures are startling. Dukes calculates that an average US gallon of gasoline required approximately 90 tonnes of precursor plant material in the process of its formation in ancient oceans (think of that every time you fill your tank). Calculated globally, human society consumes the equivalent of 400 years' worth of ancient solar energy (expressed in terms of the net primary productivity of plants during previous geological eras) each year through our use of fossil fuels. This suggests that a drastic curtailment of energy use would be necessary for humans to try and live within the current year-on-year budget of solar energy.

It also highlights the extent to which our civilisation depends on a one-off energy subsidy from the past, in the form of fossil fuels formed by plants photosynthesising ancient sunlight over millions of years. Indeed, we probably use a million years' worth of fossil fuels every year, expressed in terms of the time it took for them to form, at current rates of use. Unique amongst animals, we humans have been able to break out of the ecological constraints imposed by a finite annual budget of solar energy by using the chemical energy stored geologically in oil, coal and gas. We no longer have to rely on nature's annual bounty, because to all intents and purposes we can eat oil.

So fossil fuels, more than any other factor, have made humans successful. Other animals have to live within the constraints of their ecosystems, where their numbers are regulated by food supply, predators and so on. *Homo sapiens* has broken out from this ecological straitjacket – our food supply is no longer limited by what we can grow in the field and forage in the forest. Instead, we turn fossil fuels into food, through mechanised agriculture and long-distance transport. Natural gas is used to make nitrogen fertilisers, whilst oil powers the tractors and combine harvesters that carry out most of the labour. It would take hundreds of cider-drinking human harvesters using muscle power to do the work of one oil-drinking combine harvester using mechanical power (cider was surely the ultimate sustainable biofuel!).

Still more oil is used in processing raw materials into edible foodstuffs, packaging them, and trucking the finished products to market. The system by its nature uses vastly more energy than a pre-industrial one, and is also very inefficient: far more calories of energy from fossil fuels are put in than we get out as calories of food. We'd do better – if it were possible – just to eat the oil directly. For example, it takes 127 calories of fuel to fly in each calorie of iceberg lettuce from the United States to the UK. According to one estimate, the US food system consumes ten times more fossil energy than it produces in food energy.

With this big fossil energy subsidy, human impact on the planet and natural ecosystems has been profound. Humans already appropriate between a quarter and 40 per cent of planetary net primary production (NPP: defined as the net amount of solar energy converted to plant organic matter through photosynthesis). As the authors of one scientific study remark: 'This is a remarkable level of co-option for a species that represents roughly 0.5% of the heterotroph [animal] biomass on Earth.' And remember, this is *in addition* to the 400 years' worth of ancient NPP we consume each year through our fossil fuel use. We are voraciously consuming not just modern nature, but ancient nature too.

But fossil energy in some ways may actually have lessened the direct human impact. Instead of burning wood for heating, for example, Russians discovered that they could burn coal, reducing the pressure on forests. In the days before fossil fuels, Sweden, for example, had far fewer forests than it does now. It is highly unlikely that much of North America or Eurasia would still have significant forest cover were it not for the fossil energy subsidy. Shifting from coal to wood biomass, therefore, although desirable in terms of greenhouse gas emissions, would by this measure end up displacing even more of nature than we already have.

States of denial

Energy realities are not the only reason why our response to global warming has hitherto been so half-hearted. Our evolutionary psychology preconditions us not to respond to threats which can be postponed until later. We are good at mobilising for immediate battles, less good at heading off challenges which still lie far into the future. Hence the most appropriate term to describe both individual and societal responses so far is probably 'denial'. This is the same mental faculty that smokers use to pretend to themselves that they won't die early, or that mountain climbers scaling Everest use to imagine themselves invulnerable even as they pass the frozen bodies of previous mountaineers who have died on the very same path.

This denial is complex, involving a variety of defensive responses from the familiar 'climate change is a myth' to the more understandable (but ultimately no more useful) 'but I need my car for my job'. It is of course no coincidence that the same people who are deeply wedded to high fossil fuel use – oilmen, for example –

are the ones most likely to deny the reality of climate change. As Al Gore reminds his audience during the slideshow for his film *An Inconvenient Truth*, there is nothing so difficult as trying to get a man to understand something when his salary depends on him not understanding it. This is classic denial: no one wants to hold a mental image of themselves as bad or evil, so immoral acts are necessarily dressed up in a cloak of intellectual self-justification.

According to psychologists, denial is a way for people to resolve the dissonance caused by new information which may challenge deeply held views or cherished patterns of behaviour. Motorists, therefore, may not be willing to absorb information which challenges their perceived need to use their cars; nor are holidaymakers likely to be eager to think too much about global warming as they board their flights to Thailand. This has important implications for campaigners and educators: the denial response means that simply giving people more facts about climate change may not necessarily make them determined to act against it in any straightforward cause-and-effect way.

So why is denial easier for people than being more honest and changing their behaviour? Part of the problem is societal: we are confronted with daily social pressure to conform to a high-fossil-fuel-consuming lifestyle, so personal behavioural change in reality requires a lot of courage. Those who make the effort are frequently dismissed as 'tree-huggers' or 'sandal-wearers' by the mainstream. A high-energy lifestyle is often seen as a badge of social success.

TV and cinema adverts seek to establish high-performance cars as status symbols, for example, whilst professional people may boast about how much international travel they do. We all need validation from our peers, and if our peer group behaves in a way which undermines our beliefs about climate change, this might lead to feelings of alienation rather than satisfaction.

Given that resolving dissonance is difficult, and that denying it is dishonest, many people choose another way out of the dilemma: displacement. In short, they blame someone else. For an ordinary person this might mean singling out someone whose behaviour is worse – the Mini driver pointing to the Hummer driver, for example. For policymakers, this might mean blaming entire countries: the Byrd-Hagel resolution in the US Senate refused to countenance any change to American lifestyles unless developing countries also cut back their emissions. (In effect, it was the US blaming China.) Even environmentalists can be tempted by displacement: the vilification of George Bush – indefensible though his stance might be – is easier for most of us than having to face more tricky challenges closer to home.

Climate change is a classic 'tragedy of the commons' problem, where behaviour which makes sense at an individual level ultimately proves disastrous to society when repeated by everyone. The concept's originator, Garrett Hardin, gives the example of cattle herders using a shared pasture to illustrate the problem. Each herder stands to gain individually by adding another cow to the common – he gets more milk and beef. But if all herders act the same way, the result is overgrazing and the destruction of the shared resource. Psychological denial is integral to the process, Hardin writes: 'The individual benefits as an individual from his ability to deny the truth, even though society as a whole, of which he is a part, suffers'.

One intriguing study on this issue used random-sample focus groups in Switzerland to investigate attitudes to climate change amongst the general public. Its results showed clearly how the 'tragedy of the commons' is reflected in people's belief 'in the insignificance of individual action to change the order of things', with the result that perceived 'costs to the self are greater than benefits to others'. However, the researchers found that the most powerful motivator of denial was more straightforwardly selfish – an unwillingness to abandon personal comforts and consumption patterns. People would complain that public transport is late, dirty and overcrowded, therefore they 'need' their cars. Or they might argue that their lives are busy and difficult, so they 'need' foreign holidays for a couple of weeks a year. All these excuses seek to justify continued behaviour which is, when expressed in collective terms, highly destructive.

The study reports on a variety of other ways that denial occurs. There is the 'metaphor of displaced commitment' ('I protect the environment in other ways, like recycling'); denial of responsibility ('I am not the main cause of this problem'); condemning the accuser ('You have no right to challenge me'); rejection of blame ('I've done nothing wrong'); ignorance ('I don't know the consequences of my actions'); powerlessness ('Nothing I do makes much difference'); comfort ('It is too difficult for me to change my behaviour'); and 'fabricated constraints' ('There are too many impediments'). It is quite a list – and probably a familiar one to anyone who has discussed climate change with other people. I have heard all of these objections, in different ways, probably hundreds of times.

Perhaps the most pervasive and enduring form of denial is what the Swiss researchers call 'the faith in some form of managerial fix' – in particular the belief that the white knight of technology will come riding to the rescue. Like other forms of denial, the faith in a 'techno-fix' evades the need for any serious behavioural change. Politicians, for example, celebrate the latest prototype hydrogen car because it allows them to shift attention away from the rather more common fossil-fuelled versions. Most people believe that tackling climate change is simply a case of building enough wind turbines, fitting solar panels to enough roofs or recycling more of their glass bottles. Yet the calculations of Jeffrey Dukes, highlighting the raw figures of fossil energy use, indicate that the reality is somewhat different.

In a wider sense, one could argue that the whole economic system of modern Western society is founded on denial – in particular the denial of resource limitations. Schoolchildren are taught – and Nobel-winning economics professors apparently still believe – that Earth-provided resources, from iron ore to fisheries, come into the category of 'free goods', appearing as if by magic at the start of the economic process. These 'free goods', which include all the ecosystem services which support the human species, are considered financially valueless and missed out from conventional economic accounting.

The standard 'gross domestic product' (GDP) measuring stick of national economic success tots up the value of production and consumption without considering the sustainability of the process. In a master stroke of creative accounting, conventional economic theory therefore counts the depletion of resources as an accumulation of wealth. This is analogous to an individual spending all of the money in their current account and counting it as 'income' – an absurdity, but one which underpins our entire economy.

Bearing this societal dysfunction in mind, it is perhaps rather unfair to blame individuals for not facing up to climate change when the whole weight of economy and society works effectively in preventing them from doing so. Bob Dylan once sang about how the white Southerner who shot the black civil rights leader Medgar Evers in 1963 was 'just a pawn in their game'. So are we all, pawns in the game of global warming. But we are not entirely powerless, nor entirely blameless. The collective hand that moves these pawns is our own.

Peak oil

We may not have the luxury of choosing whether to give up fossil fuels voluntarily, however. In the last few years an increasing number of knowledgeable people have come to the conclusion that world oil supplies are close to peaking, raising the spectre of an energy crunch which would cause untold hardship. There are good reasons why they could be right. The discovery of new oil reserves has been on a steady downward trend since the mid-1960s. In 1980 the production and discovery lines on the graph crossed over – since then we have consumed more oil every year than we have discovered. Some analysts suggest that the true oil peak may come as early as next year, or may even have been passed already. No one can tell for sure, because by definition, we will only be able to see the peak in hindsight.

Part of the reason why lies in simple geology: oil is actually extremely rare. In order for an oilfield to form, a range of unlikely conditions must be fulfilled. First, large quantities of carbon must accumulate on an anoxic seabed. Dead plankton and other organic matter rains down onto the ocean floor continually, but for the vast majority of the time this carbon is consumed and oxidised by bacteria. Ocean anoxia – as we saw during the discussion on the Cretaceous period – is both rare and transient. Second, this carbon-rich sediment must form and compact in such a way that it squeezes out the water but remains permeable, with enough pore space for oil to leach through it and collect into a reservoir of porous rock above.

Third, this permeable layer must be capped with an impermeable lid (the 'cap rock') to stop the oil leaking to the surface and disappearing. Fourth, it must be buried to just the right depth to get 'cooked' by geothermal heat at the right temperature. If the temperature is too low, then the carbon stays put. If it is too high, then gas is formed instead of oil. Fifth and finally, this unlikely combination must be folded into a saddle-shaped 'anticline' (like an inverted 'U') which traps the oil in a pool under the cap rock rather than allowing it to escape, just like a pocket of air gets trapped on the ceiling of a flooded cave. Only if all these conditions are fulfilled are the long-dead bodies of plankton likely to be discovered by us millions of years later as sweet, light crude.

By far the largest field in the world is the Saudi 'super-elephant' called Ghawar, which produces an astounding 5 million barrels of oil per *day*, accounting for half of Saudi Arabia's total production. In Ghawar all the unlikely events outlined above came seamlessly together in what the American Association of Petroleum Geologists breathlessly calls a 'Geology 101 scenario': organic-rich mudstones and highly permeable grainstone form the source rock and oil reservoir, both of which are topped in turn by an impermeable cap rock of salts to keep the lid firmly on. The whole field forms an inverted 'U', at exactly the right cooking depth. 'It's basic geology,' says Abdulkader Afifi of Saudi Aramco. 'You need five conditions to form a large oil accumulation, and these things came together in a beautiful manner over a large area.'

But Ghawar may already have peaked (the Saudis aren't saying), and is known to be kept at high production rates only by the injection of large quantities of seawater to force up the remaining oil. Pretty much the entire world has now been geologically surveyed, so the chances of petroleum explorers having missed a formation like Ghawar somewhere else are negligible. With current reserves being depleted without replacement by new reserves, the 'peak oil' crowd would indeed seem to be on to something.

The British climate change campaigner and ex-geologist Jeremy Leggett warns that a failure to face up to peak oil could cause a global economic crash, combined with an upsurge in military conflict in the Middle East over the remaining oil reserves – conflict of which the US war in Iraq could be a foretaste. The energy analyst Richard Heinberg calls for a strategy he calls 'powerdown', where the world undertakes a conscious shift away from the high-energy society in order to avoid collapse on the day the oil wells begin to run dry.

There is a confusing overlap between the peak oil and climate change issues. Logically, the decline of oil supplies must be a good thing for the stability of the climate, because it would force a transition away from fossil fuels – a transition which seems unlikely to be undertaken voluntarily. In addition, high energy prices make people behave more efficiently in the way they use energy, thereby reducing emissions. High oil prices also make renewable energy more competitive, spurring further investments in solar and wind.

But fossil fuels are not only oil. Coal, still used to generate most of the world's electricity, is a larger contributor to greenhouse gas emissions than oil – and there's enough coal in the world to last another couple of centuries at least, busting any reasonable emissions budget many times over. Coal can also be turned into synthetic fuels – a technique pioneered by the Nazis and later continued by the apartheid regime in South Africa. 'Synfuels' still supply half of South Africa's petrol and diesel, and with high world oil prices, coal 'synfuels' are becoming competitive elsewhere too. The liquidation of coal produces much more CO_2 than conventional refining – so peak oil in this case would worsen global warming.

Other 'unconventional oil' sources are similarly polluting – the extraction of oil from the tar sand deposits in the Canadian province of Alberta uses vast quantities of steam and natural gas, meaning that the 'energy returned on energy invested' ratio is dangerously low and emissions dangerously high. Global gas supplies will last longer than oil, but not indefinitely – estimates of the date for 'peak gas' vary from one to eight decades away from now. But by the time conventional gas declines globally, it is very likely that the energy companies will have begun exploiting methane hydrates on ocean shelves – already a potential source of considerable interest to the likes of Exxon-Mobil and Texaco.

The picture is complicated, but it seems unlikely that peak oil will save us from global warming: even if cheap oil does indeed begin to run out sooner rather than later, the world is a long way from running short of hydrocarbons. More's the pity.

Knocking in wedges

In many countries a fractious debate has erupted about what sources of energy are best placed to replace fossil fuels. Most people quickly pin their flags to one or other mast. Greens by and large loathe nuclear power, so tend to plump for renewable options like solar and wind. Others express an equal loathing for wind turbines, and mount vociferous campaigns against their development. Some take both or neither side. In Cape Cod, Massachusetts, prominent environmental activists like Robert F. Kennedy Jr have opposed a large offshore wind farm, despite its potential for clean power generation – because, some suspect, it would spoil their view. In the UK, the scientist James Lovelock is at least consistent; though he fulminates against wind turbines, he is a passionate advocate of nuclear power.

With each side offering its miracle energy cure, the public is left with a false impression – that we simply have to choose one of the touted solutions and the problem will be solved. The reality is that only a combination of serious energy efficiency and a wide variety of new technologies offers any hope of a way out of the crisis. This essential truth was illustrated very usefully a couple of years ago by Robert Socolow and Steve Pacala of Princeton University in

New Jersey. Their idea was to consider each technology a potential 'wedge' – one of a number of wedges which together would make the difference between an upwards emissions trend or a stabilising one. Each wedge represents the reduction of annual emissions by one billion tonnes of carbon by the year 2055, and implementing seven wedges will enable the world to achieve the goal of emitting no more CO_2 in 2055 than today. Socolow and Pacala urged people not to be 'beguiled by the possibility of revolutionary technology' like nuclear fusion, artificial photosynthesis or space-based solar electricity. Instead, they emphasised, 'humanity can solve the carbon and climate problem in the first half of this century simply by scaling up what we already know how to do'.

Socolow and Pacala's paper confirms beyond doubt that there is no 'silver bullet' which on its own will allow humanity to kick the carbon habit whilst continuing each year to use more energy. Whilst one wedge can be gained by increasing the world vehicle fleet's fuel economy from 30 miles per gallon to 60 mpg, for example, the same carbon displacement could also be achieved by halving the average distance travelled per car from 10,000 miles per year to 5,000. More efficient buildings and electricity generation can also deliver a wedge each. If gas displaced coal in electricity generation, a quadrupling of gas-fuelled power stations gives another wedge. An additional 700 one-gigawatt nuclear power stations worldwide could displace enough coal for another power-generation wedge, as could introducing 'carbon capture and storage' (the pumping of CO_2 captured from chimneys underground into geological reservoirs) at 800 similar-sized coal plants. Remember, we need seven of these wedges just to stabilise global emissions at their current levels.

Socolow and Pacala's approach is especially useful because it illustrates the importance of scale. This is particularly daunting for renewables – for wind power to achieve one wedge, 2 million

1-megawatt turbines would be needed, a fifty-fold increase from today's deployment. The turbines would cover 30 million hectares, equivalent to 3 per cent of the total land area of the United States. A wedge of solar photovoltaic electricity generation would need a 700-fold increase from today's total, covering 2 million hectares of land – or around 3 square metres per person. Wind-generated power could electrolyse water, producing clean hydrogen for fuel-cell-driven cars – but this would require another 4 million 1 MW wind turbines to displace a wedge-worth of petrol and diesel fuel. A massive programme of reforestation, combined with an end to the clear-cutting of tropical forests, might also deliver a wedge of carbon emissions reductions.

All of these approaches have their pros and cons, of course. Wind turbines kill birds, though this effect can be reduced by sensitive siting decisions, away from migration paths and areas frequented by birds of prey like golden eagles. It also tends to be exaggerated by the antis: one US-wide survey suggests that 40,000 birds perish in turbine blades every year, which sounds a lot until contrasted with the number killed by domestic cats, currently estimated to be in the hundreds of millions. To be fully consistent, therefore, anti-wind farm campaigners concerned about dangers to bird populations would probably be better occupied grabbing a shotgun and conducting a cull of the local neighbourhood moggies. Wind power is also a potentially enormous energy source: one 2007 study suggests that if the continental shelf offshore from the eastern United States were planted with 166,000 5 MW turbines, the resulting electricity could power the entire east coast region, from Massachusetts to North Carolina. Fossil fuels could be displaced not only from existing coal-fired power stations, but from vehicles and buildings too – if plug-in hybrids or fuel-celled vehicles, ground source heat pumps and other electrical substitutes were taken up widely. Add this together, and nearly 70 per cent of the region's CO_2 emissions could be wiped out permanently with this one single energy source – a cut which, if rolled out more widely, would go most of the way to stabilising the world's climate. All that is lacking is political will and investment.

Unlike wind turbines, which allow agriculture or fisheries to continue underneath them more or less unhindered, land devoted to solar panels would be unusable for anything else; although I can't think of any obvious drawbacks to putting solar panels on roofs. Some have suggested that covering large areas of the world's deserts with solar photovoltaic panels would be a good way to harvest the brightest sunshine without sacrificing much-needed farmland, particularly if the electricity generated could then be transported vast distances to high-energy-consuming countries via low-resistance DC cables. Even if such colossal-scale construction efforts could realistically be undertaken, deserts are ecosystems too, and have a role in the planet's natural stability even if they look barren and inhospitable to the human eye. However, along with wind, there is no doubt that enough solar radiation hits the planet's surface to easily supply humanity's energy use many times over.

More controversial than either of these renewable sources is nuclear power, a very low-carbon energy source, and one with a proven track record in electricity generation – but one which also raises dangers of nuclear weapons proliferation and deadly accidents, as well as the still unsolved question of what to do with highly radioactive wastes. Carbon capture and storage, the piping of liquid CO_2 underground into deep saline aquifers or old oil wells, is an unproven technology which could result in unexpected releases of CO_2 from faulty or leaky reservoirs below ground, although the IPCC ranks this risk as very low. Energy efficiency measures have fewer drawbacks – but improving efficiency in cars and buildings can have the surprising effect of increasing power consumption overall by making energy cheaper than it would otherwise be, and it is an open question to what extent large-scale energy reduction measures are really compatible with a growing economy.

Perhaps the easiest and most beneficial way to reduce emissions would be to stop the destruction of tropical forests. Although through most of this book I have focused on emissions from fossil fuel burning, estimates suggest that up to 20 per cent of human greenhouse gas emissions come from deforestation in the tropics. Brazil and Indonesia are two of the highest carbon emitters in the world – not because their inhabitants are too fond of their cars, but because their huge areas of forest are being levelled at an accelerating pace. Halting this global blitzkrieg, on the other hand, would save the same amount of carbon over the next century as stopping all fossil fuel emissions for an entire decade. Given that tropical forests are vital for biodiversity, and as carbon sinks, protecting them is clearly a win-win option. Although 'avoided deforestation' was excluded from the first round of the Kyoto Protocol (both because of sovereignty concerns from the nations involved and worries about how to properly account for it), this may change in the second phase, due to start in 2012. The only problem is money: given that the timber stored in these trees has a high value, whilst the carbon stored in them currently has none, if we in rich countries want poorer countries to stop chopping down their forests, we're going to have to pay them for it.

If avoiding deforestation is perhaps the best wedge option, the worst may be that of biofuels. Already corn-derived ethanol is being blended into gasoline in the United States, ostensibly to reduce CO_2 emissions, but in reality having more to do with subsidising the politically powerful farming lobby. It is also far from clear whether any carbon is actually displaced, given that the production, milling and transportation of corn uses large amounts of fossil fuel in trucks, tractors and factories. Some green enthusiasts already run their cars on used chip fat, and are strong advocates of biofuels. But again the scale issue is crucial. Although no one can object to using waste vegetable oils from restaurants as the feedstock for biodiesel, this source could only provide a few hundredths of one per cent of the fuel used by the entire nation-wide car fleet.

Other biofuel advocates point to waste straw or wood chippings as a way to manufacture ethanol from cellulose, perhaps using genetically engineered enzymes. This seems to hold more poten-tial in terms of carbon displacement, as it could be far more efficient than producing ethanol from food crops. However, the techniques are still being developed, and would take years to scale up enough to make any serious dent in emissions. In addition, there are still problems – if straw is not ploughed back into the land, then biomass is being removed, lowering the humus content of the soil and its nourishment value for plants. If the use of woody residues causes coppice plantations to expand into current-ly 'marginal' areas, then the wild spaces where nature currently clings on will be reduced still further. Mankind's destructive need for food will be increasingly augmented by a destructive need for energy.

Leaving aside the biodiversity issue and looking just at existing manufacturing techniques, an ethanol wedge in Socolow and Pacala's analysis would require 250 million hectares being devoted to corn or sugar cane plantations, an area equivalent to one-sixth of the world's croplands. Given that world food stocks are already at historic lows because of population growth and droughts, devoting more of our best farmland to growing fuel for cars seems close to insane. It may also be immoral: because car-owning people are by definition among the world's rich elite, using food crops to replace petrol would create scarcity and drive up food prices on the commodity markets, leaving the poorest to starve. The reality is simple: you can use land to feed cars or to feed people, but not both.

A related question arises with the European Union's target of 5 per cent biofuels in its vehicle fleet by 2010. Much of this will come from biodiesel, and a major feedstock for this is palm oil grown on plantations in Indonesia and Malaysia. These plantations have been responsible for disastrous clear-cutting of the fast-declining natural tropical forests, destroying the habitat of rare species like the orang-utan and causing major additional carbon releases through the burning of wood and underlying peat. As discussed above, these Asian forest fires are the greatest single cause of greenhouse gas emissions apart from fossil fuel use: during the 1997–8 drought, an estimated 2 billion tonnes of carbon poured out of south-east Asian forests as they burned. Some of these fires were natural, but many were also caused by plantation owners setting light to virgin forest in order to clear it. It does not therefore require much imagination to understand why 'deforestation diesel' almost certainly has a worse impact on global warming than its conventional mineral counterpart: estimates have suggested that biodiesel based on palm oil feedstock can be ten times more carbon-intensive than fossil fuels.

If we leave biofuels and nuclear out of any prospective energy portfolio because of their obvious drawbacks, we can still get our seven wedges in other ways. We need to halve the distances people drive each year, and we need to double vehicle fuel economy. We need to dramatically increase the efficiency of buildings and fossil-fuelled power stations. We need to construct 2 million 1 MW wind turbines to generate electricity, and cover 2 million hectares of land with solar panels. We need to stop the destruction of tropical forests, and we need to dramatically increase tree cover elsewhere. And we need to make a difficult choice between injecting billions of tonnes of carbon dioxide underground and investing in 1,400 new gas power plants to produce electricity.

All this, and we can hope to stabilise emissions in 2055 at today's levels, breaking the continual upward growth of a 'business as usual' path. But this still leaves us with a problem. As I mentioned earlier, simply stabilising emissions at today's levels in fifty years isn't nearly enough to keep us within the two degrees safety target. We need to *cut* emissions, and to do so within a decade. Socolow and Pacala estimate that their seven wedges take the world towards a 500 ppm trajectory, 100 ppm higher than the level I consider necessary to stabilise the climate below two degrees of global warming. Indeed, a 500 ppm concentration would probably warm the globe by between three and four degrees, perhaps crossing both the carbon cycle and Siberian methane tipping points in the process.

To stay within the 400 ppm two degrees target, on the other hand, we need to knock another four or five wedges into the emissions graph. This is not impossible – we can double the quantity of wind turbines, and further reduce the numbers of vehicles on the roads. We can cut our need for energy by living less consumptive lifestyles, and by adopting more localised patterns of behaviour. As Socolow and Pacala emphasise, we already have the technologies and the social know-how to achieve this transition. But that still leaves the political question open: is there a way to turn our society around, so that people work eagerly and collectively to achieve a two degrees target, rather than either ignoring the whole issue or trying to find ways to justify their own damaging behaviour? Currently, economic and social pressure works in the other direction: young children, for example, rather than eschewing consumerism, have to display the latest consumer fashion gear so as not to be humiliated in the playground. Their parents save up to buy the latest jeep or SUV in order to demonstrate their status and earning power to the neighbours. Television programmes equate speed with virility and driving with freedom, cultural messages which are relentlessly reinforced by screen and billboard advertising. Hip hop artists reinforce these values by rapping about their expensive cars and making videos with scantily clad models dancing sexily around them.

It seems clear to me that no amount of barracking from bearded hippie environmentalists will persuade society at large that the consumer treadmill is not the quickest route to health and happiness. Most of my neighbours still shop in supermarkets, even though they have to drive to them in cars and are depriving local shopkeepers from making a living in the process. An outdated view still prevails that a low-carbon lifestyle requires immense personal suffering and sacrifice. In my view, nothing could be further from the truth. All the evidence shows that people who do not drive, do not fly on planes, do shop locally, do grow their own food, and do get to know other members of their community have a much higher quality of life than their compatriots who remain addicted to high-fossil-fuel-consuming lifestyles.

Just as people were better off and healthier in Britain under food rationing during the Second World War, so most of us would see a dramatic improvement in our quality of life if 'carbon rationing' were introduced by the government. Such a scheme need not be technically complex or difficult to introduce: people could simply trade carbon as a parallel virtual currency, swiping their carbon cards at the petrol pump and surrendering the requisite amount of carbon ration when buying flights or paying their electricity bill. Although carbon permits should be tradeable in the interests of flexibility, conspicuous carbon consumption by celebrities would be largely eliminated. Instead, social pressure would reverse, with people happy to make changes in the knowledge that everyone else is doing likewise.

As traffic danger ceased, young children could play football in the streets again.

In constraining carbon through rationing, we might soon find that we were building a different sort of society, one emphasising quality of life before the raw statistics of economic growth and relentless consumerism. I have no grand plan for how this society might look, nor do I pretend that it would be some kind of utopia. Life would go on, with all its trials and tribulations – and that, after all, is precisely the point. Unless we do constrain carbon, life will very largely not go on at all.

It seems to me that this low-carbon society would be one which remembers that our planet is a unique gift – perhaps the only one of its kind in the entire universe – which we are indescribably privileged to be born into. It would be a society that could look back on the six degrees nightmare scenario as just that – a nightmare, one which humanity woke up from and avoided before it was too late. More than anything, it would be a society which survived and prospered, and which passed on this glorious inheritance – of ice caps, rainforests and thriving civilisations – to countless generations, far into the future.

NOTES

INTRODUCTION

p. xvii   **six degrees colder:** Schneider von Deimling, T., et al., 2006: 'How cold was the Last Glacial Maximum?', *Geophysical Research Letters*, 33, 14, L14709. The researchers conclude that global mean cooling relative to pre-industrial climate (i.e. 0.8°C cooler than ours) was 5.8°C ± 1.4°C, substantially colder than previous model-based estimates.

p. xviii   **human population crashed:** Burroughs, W., 2005: *Climate Change in Prehistory: The End of the Reign of Chaos*, Cambridge University Press, p. 139

1 ONE DEGREE

p. 3   **stumps dated from medieval times:** Stine, S., 1994: 'Extreme and persistent drought in California and Patagonia during mediaeval time', *Nature*, 369, 546–9

p. 4   **same two time intervals:** Ibid.

p. 4   **tinder-dry:** Swetnam, T., 1993: 'Fire history and climate change in giant sequoia groves', *Science*, 262, 885–9

p. 5   **epic droughts:** Laird, K., et al., 1996: 'Greater drought intensity and frequency before AD 1200 in the Northern Great Plains, USA', *Nature*, 384, 552–4

p. 5 **flash floods:** Meyer, G., and Pierce, J., 2003: 'Climatic controls on fire-induced sediment pulses in Yellowstone National Park and central Idaho: a long-term perspective', *Forest Ecology and Management*, 178, 1–2, 89–104

p. 6 **violent conflict:** Diamond, J., 2005: *Collapse: How Societies Choose to Fail or Survive*, Allen Lane

p. 6 **warming and then cooling:** Jones, P., and Mann, M., 2004: 'Climate over past millennia', *Reviews of Geophysics*, 42, RG2002

p. 6 **medieval flows:** Meko, D., et al., 2007: 'Medieval drought in the upper Colorado River Basin', *Geophysical Research Letters*, 34, L10705

p. 8 **drought ... over decades:** Mangan, J., et al., 2004: 'Response of Nebraska Sand Hills natural vegetation to drought, fire, grazing, and plant functional type shifts as simulated by the century model', *Climatic Change*, 63, 49–90

p. 10 **cooling also occurred:** Committee on Abrupt Climate Change, 2002: *Abrupt Climate Change: Inevitable Surprises*, chapter 2 – Evidence of Abrupt Climate Change

p. 10 **surge of water:** Burroughs, W., 2005: *Climate Change in Prehistory: The End of the Reign of Chaos*, Cambridge University Press, p. 61

p. 11 **circulation had dropped:** Bryden, H., et al., 2005: 'Slowing of the Atlantic meridional overturning circulation at 25°N', *Nature*, 438, 655–7

p. 12 **winters ... of 1962–63:** 'Great weather events: the winter of 1962/63', UK Met Office, http://www.metoffice.com/corporate/pressoffice/anniversary/winter1962-63.html

p. 12 **50 per cent drop:** Jacob, D., et al., 2005: 'Slowdown of the thermohaline circulation causes enhanced maritime climate influence and snow cover over Europe', *Geophysical Research Letters*, 32, L21711

p. 12 **random natural variability**: Kerr, R., 2006: 'False alarm: Atlantic conveyor belt hasn't slowed down after all', *Science*, 314, 1064

p. 13 **'still warming'**: Climate Change 2007: The Physical Science Basis. Contribution of Working Group I to the Fourth Assessment Report of the Intergovernmental Panel on Climate Change – chapter 10: Global Climate Projections, Executive Summary

p. 14 **Thompson … camped**: Krajik, K., 2002: 'Ice man: Lonnie Thompson scales the peaks for science', *Science*, 298, 518–22

p. 15 **ice had already melted**: Thompson, L., et al., 2002: 'Kilimanjaro ice core records: Evidence of Holocene climate change in tropical Africa', *Science*, 298, 589–93

p. 15 **'ice will disappear'**: 2001: 'Deciphering the ice', CNN, http://edition.cnn.com/SPECIALS/2001/americasbest/ science.medicine/pro.lthompson.html

p. 17 **glacial retreat in the Rwenzoris**: Taylor, R. G., et al., 2006: 'Recent glacial recession in the Rwenzori Mountains of East Africa due to rising air temperature', *Geophysical Research Letters*, 33, 10, L10402

p. 17 **15 million cubic metres**: Agrawala, S., et al., 2003: 'Development and climate change in Tanzania: Focus on Mount Kilimanjaro', OECD Environment Directorate, 67pp

p. 19 **Neolithic paintings**: Brooks, N., et al., undated: 'The prehistory of Western Sahara in a regional context', http://www.uea.ac.uk/sahara/publications/nb_west_abs.pdf

p. 19 **arrowheads and flint knives**: Kindermann, K., et al., 2006: 'Palaeoenvironment and Holocene land use of Djara, Western Desert of Egypt', *Quaternary Science Reviews*, 25, 13–14, 1619–1637

p. 19 **flint fish-hooks:** Fezzan Project – Archaeology, http://www.cru.uea.ac.uk/%7Ee118/Fezzan/fezzan_archaeol.html

p. 19 **river valleys buried:** NASA Earth Observatory Newsroom, http://earthobservatory.nasa.gov/Newsroom/NewImages/images.php3?img_id=16963

p. 20 **desert edge retreated:** Gasse, F., 2002: 'Diatom-inferred salinity and carbonate oxygen isotopes in Holocene waterbodies of the western Sahara and Sahel (Africa)', *Quaternary Science Reviews*, 21, 737–67

p. 20 **largest freshwater body:** Leblanc, M., et al., 2006: 'Reconstruction of Megalake Chad using Shuttle Radar Topographic Mission data', *Palaeogeography, Palaeoclimatology, Palaeoecology*, 239, 1–2, 16–27

p. 20 **warm, shallow waters:** Schuster, M., et al., 2005: 'Holocene Lake Mega-Chad palaeoshorelines from space', *Quaternary Science Reviews*, 24, 1821–7

p. 20 **erosive power:** Leblanc, M., et al., 2006: 'Reconstruction of Megalake Chad using Shuttle Radar Topographic Mission data', *Palaeogeography, Palaeoclimatology, Palaeoecology*, 239, 1–2, 16–27

p. 20 **powered a monsoon:** Claussen, M., et al., 1999: 'Simulation of an abrupt change in Saharan vegetation in the mid-Holocene', *Geophysical Research Letters*, 26, 14, 2037–40

p. 21 **long droughts:** Kindermann, K., et al., 2006: 'Palaeoenvironment and Holocene land use of Djara, Western Desert of Egypt', *Quaternary Science Reviews*, 25, 13–14, 1619–1637

p. 21 **get wetter:** Hoerling, M., et al., 2005: 'Detection and attribution of 20th century northern and southern African monsoon change', *Journal of Climate*, 19, 16, 3989–4008

p. 22 **heavier rains:** Paeth, H., and Hense, A., 2004: 'SST versus climate change signals in West African rainfall: 20th century variations and future projections', *Climatic Change*, 65, 179–208

p. 22 **even fiercer drought:** Held, I., et al., 2005: 'Simulation of Sahel drought in the 20th and 21st centuries', *Proceedings of the National Academy of Sciences*, 102, 50, 17891–6

p. 24 **temperatures in Greenland:** Climate Change 2007: The Physical Science Basis. Contribution of Working Group I to the Fourth Assessment Report of the Intergovernmental Panel on Climate Change – chapter 6: Palaeoclimate, section 6.4.2

p. 24 **as warm as now:** Climate Change 2007: The Physical Science Basis. Contribution of Working Group I to the Fourth Assessment Report of the Intergovernmental Panel on Climate Change – chapter 6: Palaeoclimate, section 6.6.1

p. 24 **within a degree:** Hansen, J., et al., 2006: 'Global temperature change', *Proceedings of the National Academy of Sciences*, 103, 39, 14288–93

p. 25 **risen by 2–3°C:** Arctic Climate Impact Assessment, 2004: *Impacts of a Warming Arctic*, Cambridge University Press

p. 25 **sudden thawing:** Torre Jorgenson, M., et al., 2006: 'Abrupt increase in permafrost degradation in Arctic Alaska', *Geophysical Research Letters*, 33, L02503

p. 25 **shrunk or disappeared:** Riordan, B., et al., 2006: 'Shrinking ponds in subarctic Alaska based on 1950–2002 remotely sensed images', *Journal of Geophysical Research*, 111, G04002

p. 25 **catching fire:** Smol, J., and Douglas, M., 2007: 'Crossing the final ecological threshold in high Arctic ponds', *Proceedings of the National Academy of Sciences*, 104, 30, 12395–7

p. 25 **global sea levels:** Arendt, A., et al., 2002: 'Rapid wastage of Alaska glaciers and their contribution to rising sea level', *Science*, 297, 382–6

p. 26 **lost 400 cubic kilometres:** Hinzman, L., et al., 2005: 'Evidence and implications of recent climate change in Northern Alaska and other Arctic regions', *Climatic Change*, 72, 251–98

p. 26 **vanished:** Serreze, M., Holland, M., and Stroeve, J., 2007: 'Perspectives on the Arctic's shrinking sea-ice cover', *Science*, 315, 1533–6

p. 26 **ice extent:** Comiso, J., 2006: 'Abrupt decline in the Arctic winter sea ice cover', *Geophysical Research Letters*, 33, L18504

p. 26 **virtually ice-free:** Holland, M., Bitz, C., and Tremblay, B., 2007: 'Future abrupt reductions in the summer Arctic sea ice', *Geophysical Research Letters*, 33, L23503

p. 27 **'chance of avoiding':** Hansen, J., et al., 2007: 'Climate change and trace gases', *Philosophical Transactions of the Royal Society A*, 365, 1925–54

p. 27 **ice extent reductions:** Reported on http://arctic.atmos.uiuc.edu/cryosphere/, for 9 August 2007

p. 27 **tipping point:** Foley, J., 2005: 'Tipping points in the tundra', *Science*, 310, 627–8

p. 28 **latitudinal contraction:** Fu, Q., et al., 2006: 'Enhanced mid-latitude tropospheric warming in satellite measurements', *Science*, 312, 1179

p. 28 **'outside the envelope':** Quoted in Foley, J., 2005: 'Tipping points in the tundra', *Science*, 310, 627–8

p. 30 **summer of 2003:** Schiermeier, Q., 2003: 'Alpine thaw breaks ice over permafrost's role', *Nature*, 424, 712

p. 30 **scientific paper:** Gruber, S., Hoezle, M., and Haeberli, W., 2004: 'Permafrost thaw and destabilisation of Alpine rock walls in the hot summer of 2003', *Geophysical Research Letters*, 31, L13504

p. 31 **Wet Tropics:** General information at http://www.wettropics. gov.au/pa/pa_default.html and subsequent pages

p. 33 **above certain heights:** 2002: 'The future of life in the rainforest canopy', Earthbeat, Radio National, http://www.abc.net.au/rn/science/earth/stories/s601197.htm

p. 33 **'environmental catastrophe':** Williams, S., et al., 2003: 'Climate change in Australian tropical rainforests: an impending environmental catastrophe', *Proceedings of the Royal Society of London B*, 270, 1887–92

p. 33 **highland rainforest:** Hilbert, D., et al., 2001: 'Sensitivity of tropical forests to climate change in the humid tropics of North Queensland', *Austral Ecology*, 26, 590–603

p. 35 **past millennia:** Hoegh-Guldberg, O., personal communication

p. 35 **60–95 per cent:** 2003: 'Securing Australia's Great Barrier Reef', WWF Australia, http://www.wwf.org.au/News_and_ information/Publications/PDF/Policies_position/securing_ australias_great_barrier_reef.pdf

p. 36 **landmark 1999 paper:** Hoegh-Guldberg, O., 1999: 'Climate change, coral bleaching and the future of the world's coral reefs', *Marine and Freshwater Research*, 50, 839–66

p. 37 **heat-tolerant:** Baker, A., et al., 2004: 'Corals' adaptive response to climate change', *Nature*, 430, 741

p. 37 **Severe bleaching:** Donner, S., et al., 2005: 'Global assessment of coral bleaching and required rates of adaptation under climate change', *Global Change Biology*, 11, 2251–65

p. 38 **every two years:** Donner, S., Knutson, T., and Oppenheimer, M., 2007: 'Model-based assessment of the role of human-induced climate change in the 2005 Caribbean coral bleaching event', *Proceedings of the*

National Academy of Sciences, 104, 13, 5483–8

p. 38 **a third:** Stone, R., 2007: 'A world without corals?', *Science*, 316, 678–81

p. 39 **Cape Floristic Region:** See Conservation International's outline on Biodiversity Hotspots, available at http://www.biodiversityhotspots.org/xp/Hotspots/cape_floristic/biodiversity.xml

p. 39 **devastating impact:** Bomhard, B., et al., 2005: 'Potential impacts of future land use and climate change on the Red List status of the Proteaceae in the Cape Floristic region, South Africa', *Global Change Biology*, 11, 1452–68

p. 39 **pikas:** See WWF information on threatened species: pikas, at http://www.panda.org/about_wwf/what_we_do/climate_change/problems/impacts/species/pikas/index.cfm

p. 40 **Monteverde Cloud Forest:** 'While still verdant enough to justify its name, the Monteverde Cloud Forest Preserve is beginning to resemble a crown that has lost its brightest and most beautiful gems,' writes Tim Flannery. I am indebted to him for his extensive quotations of Marty Crump's work. See Flannery, T., 2005: *The Weather Makers*, Allen Lane, chapter 12.

p. 41 **harlequin frog:** Pounds, A., et al., 2006: 'Widespread amphibian extinctions from epidemic disease driven by global warming', *Nature*, 439, 161–7

p. 41 **central:** 'Global warming and amphibian losses', three articles published in *Nature*, 447, 31 May 2007, E3–E6

p. 42 **Hurricane Wilma:** Wilma's minimum central pressure dropped to 882 mb on 19 October 2005, whilst wind speeds rose to 175 mph. See report at http://www.nhc.noaa.gov/archive/2005/tws/MIATWSAT_oct.shtml

p. 43 **forensic dissection of Catarina:** Bernardes Pezza, A., and Simmonds, I., 2005: 'The first South Atlantic hurricane:

Unprecedented blocking, low shear and climate change', *Geophysical Research Letters*, 32, L15712

p. 45 **symmetrical eye:** Gaertner, M., et al., 2007: 'Tropical cyclones over the Mediterranean Sea in climate change simulations', *Geophysical Research Letters*, 34, L14711

p. 45 **academic discussion:** See for example William Gray's response, submitted to *Nature*: http://tropical.atmos. colostate.edu/Includes/Documents/Responses/emanuel_comments.pdf

p. 45 **second piece:** Webster, P., et al., 2005: 'Changes in tropical cyclone number, duration and intensity in a warming environment', *Science*, 309, 1844–6

p. 46 **at least half:** Trenberth, K., and Shea, D., 2006: 'Atlantic hurricanes and natural variability in 2005', *Geophysical Research Letters*, 33, L12704

p. 47 **begin to unravel:** Arenstam Gibbons, S., and Nicholls, R., 2006: 'Island abandonment and sea level rise: An historical analog from the Chesapeake Bay, USA', *Global Environmental Change*, 16, 1, 40–47

2 TWO DEGREES

p. 52 **Chinese scientists:** Chen, F., et al., 2003: 'Stable East Asian monsoon climate during the Last Interglacial (Eemian) indicated by paleosol S1 in the western part of the Chinese Loess Plateau', *Global and Planetary Change*, 36, 171–9

p. 52 **about 1°C higher:** Hansen, J., et al., 2006: 'Global temperature change', *Proceedings of the National Academy of Sciences*, 103, 39, 14288–93

p. 54 **reduce the alkalinity:** Orr, J., et al., 2005: 'Anthropogenic ocean acidification over the twenty-first century and its impact on calcifying organisms', *Nature*, 437, 681–6

p. 54 **major report:** The Royal Society, 2005: *Ocean acidification due to increasing atmospheric carbon dioxide*, Policy Document 12/05

p. 54 **toxic:** Orr, J., et al., 2005: 'Anthropogenic ocean acidification over the twenty-first century and its impact on calcifying organisms', *Nature*, 437, 681–6

p. 55 **began to disintegrate:** Ruttimann, J., 2006: 'Sick seas', *Nature*, 442, 978–80

p. 55 **will dissolve:** Gazeau, F., et al., 2007: 'Impact of elevated CO_2 on shellfish calcification', *Geophysical Research Letters*, 34, L07603

p. 55 **'huge risk':** Schiermeier, Q., 2004: 'Researchers seek to turn the tide on problem of acid seas', *Nature*, 430, 820

p. 56 **decline in plankton:** Behrenfeld, M., et al., 2006: 'Climate-driven trends in contemporary ocean productivity', *Nature*, 444, 752–5

p. 57 **Switzerland:** Beniston, M., and Diaz, H., 2004: 'The 2003 heat wave as an example of summers in a greenhouse climate? Observations and climate model simulations for Basel, Switzerland', *Global and Planetary Change*, 44, 73–81

p. 58 **forest fires:** Schar, C., and Jendritzky, G., 2004: 'Hot news from summer 2003', *Nature*, 432, 559–60

p. 58 **Melt rates:** World Meteorological Institute: see http://www.wmo.int/web/Press/Press702_en.doc

p. 58 **melting permafrost:** Beniston, M., and Diaz, H., 2004: 'The 2003 heat wave as an example of summers in a greenhouse climate? Observations and climate model simulations for Basel, Switzerland', *Global and Planetary Change*, 44, 73–81

p. 59 **off the statistical scale:** Schar, C., et al., 2004: 'The role of increasing temperature variability in summer heatwaves', *Nature*, 427, 332–6

p. 59 **doubled the risk:** Stott, P. , Stone, D., and Allen, M., 2004: 'Human contribution to the European heatwave of 2003', *Nature*, 432, 610–14

p. 59 **across Europe:** Della-Marta, P. , et al., 2007: 'Doubled length of western European summer heat waves since 1880', *Journal of Geophysical Research (Atmospheres)*, 112, D15103

p. 59 **warmer than 2003:** Stott, P., Stone, D., and Allen, M., 2004: 'Human contribution to the European heatwave of 2003', *Nature*, 432, 610-14

p. 60 **drop in plant growth:** Ciais, Ph., et al., 2005: 'Europe-wide reduction in primary productivity caused by the heat and drought in 2003', *Nature*, 437, 529–33

p. 60 **European plants:** Ibid.

p. 60 **accumulated in the atmosphere:** Zeng, N., and Haifeng, Q., 2005: 'Impact of the 1998–2002 midlatitude drought and warming on terrestrial ecosystem and the global carbon cycle', *Geophysical Research Letters*, 32, L22709

p. 61 **visible from space:** NASA pictures are available at http://earthobservatory.nasa.gov/NaturalHazards/natural_hazards_v2.php3?img_id=11709

p. 61 **increasingly common sight:** Giannakopolous, C., et al., 2005: 'Climate change impacts in the Mediterranean resulting from a 2°C global temperature rise', WWF, July 2005

p. 63 **many metres higher:** Vezina, J., Jones, B., and Ford D., 1999: 'Sea-level highstands over the last 500,000 years; evidence from the Ironshore Formation on Grand Cayman, British West Indies', *Journal of Sedimentary Research*, 69, 2, 317–27

p. 64 **5–6 metres above:** Tarasov, L., and Richard Peltier, W., 2003: 'Greenland glacial history, borehole constraints and

Eemian extent', *Journal of Geophysical Research*, 108, B3, 2143; Overpeck, J., et al., 2006: 'Paleoclimatic evidence for future ice-sheet instability and rapid sea-level rise', *Science*, 311, 1747–50

p. 64 **about 1°C higher than now:** James Hansen estimates about 1°C. See Hansen, J., 2005: 'A slippery slope: How much global warming constitutes "dangerous anthropogenic interference"?', *Climatic Change*, 68, 269–79. According to E. Rohling, et al., the difference was 2°C globally. See Rohling, E., et al., 2002: 'African monsoon variability during the previous interglacial maximum', *Earth and Planetary Science Letters*, 202, 61–75.

p. 64 **contains enough ice:** Oppenheimer, M., and Alley, R., 2004: 'The West Antarctic Ice Sheet and long term climate policy', *Climatic Change*, 64, 1–2, 1–10

p. 64 **'threat of disaster':** Mercer, J., 1978: 'West Antarctic ice sheet and CO_2 greenhouse effect: a threat of disaster', *Nature*, 271, 321–5

p. 64 **free of ice:** Cuffey, K., and Marshall, S., 2000: 'Substantial contribution to sea-level rise during the last interglacial from the Greenland ice sheet', *Nature*, 404, 591–4

p. 64 **once forested:** Willerslev, E., et al., 2007: 'Ancient biomolecules from deep ice cores reveal a forested southern Greenland', *Science*, 317, 111–14

p. 65 **between 2 and 5 metres:** Tarasov, L., and Richard Peltier, W., 2003: 'Greenland glacial history, borehole constraints and Eemian extent', *Journal of Geophysical Research*, 108, B3, 2143; Otto-Bliesner, B., et al., 2006: 'Simulating Arctic climate warmth and icefield retreat in the last interglaciation', *Science*, 311, 1751–3

p. 66 **far greater:** Hansen, J., 2005: 'A slippery slope: How much global warming constitutes "dangerous anthropogenic interference"?', *Climatic Change*, 68, 269–79

p. 66 **tropical coral reefs:** Webster, J., et al., 2004: 'Drowning of the –150 m reef off Hawaii: A casualty of global meltwater pulse 1A?', *Geology*, 32, 3, 249–52

p. 66 **submerging:** Kienast, M., et al., 2003: 'Synchroneity of meltwater pulse 1A and the Bolling warming: New evidence from the South China Sea', *Geology*, 31, 1, 67–70

p. 66 **'explosively rapid':** Hansen, J., 2005: 'A slippery slope: How much global warming constitutes "dangerous anthropogenic interference"?', *Climatic Change*, 68, 269–79

p. 66 **2.7°C:** Gregory, J., Huybrechts, P., and Raper, S., 2004: 'Threatened loss of the Greenland ice sheet', *Nature*, 428, 616

p. 66 **2.2 times:** Chylek, P., and Lohmann, U., 2005: 'Ratio of the Greenland to global temperature change: Comparison to observations and climate modelling results', *Geophysical Research Letters*, 32, L14705

p. 67 **6 cm a year:** Johannessen, O., et al., 2005: 'Recent ice-sheet growth in the interior of Greenland', *Science*, 310, 1013–16

p. 67 **offset rising sea levels:** Bugnion, V., and Stone, P., 2002: 'Snowpack model estimates of the mass balance of the Greenland ice sheet and its changes over the twenty-first century', *Climate Dynamics*, 20, 87–106

p. 67 **'thinning like mad':** Schiermeier, Q., 2004: 'A rising tide', *Nature*, 428, 114–15

p. 68 **thinner ice cap:** Parizek, B., and Alley, R., 2004: 'Implications of increased Greenland surface melt under global-warming scenarios: ice-sheet simulations', *Quaternary Science Reviews*, 23, 1013–27

p. 68 **Jakobshavn Isbrae:** Joughin, I., et al., 2004: 'Large fluctuations in speed on Greenland's Jakobshavn Isbrae glacier', *Nature*, 432, 608–10

p. 68   **ice flow speeded up:** Howat, I., et al., 2005: 'Rapid retreat and acceleration of Helheim Glacier, east Greenland', *Geophysical Research Letters*, 32, L22502

p. 69   **Kangerdlugssuaq Glacier:** Luckman, A., et al., 2006: 'Rapid and synchronous ice-dynamic changes in East Greenland', *Geophysical Research Letters*, 33, L03503

p. 69   **doubled the rate:** Ibid.

p. 69   **normality:** Howat, I., et al., 2007: 'Rapid changes in ice discharge from Greenland outlet glaciers', *Science*, 315, 1559–61

p. 69   **'more vulnerable':** Truffer, M., and Fahnestock, M., 2007: 'Rethinking ice sheet time scales', *Science*, 315, 1508–10

p. 70   **100 billion tonnes:** Luthcke, S., et al., 2006: 'Recent Greenland ice mass loss by drainage system from satellite gravity observations', *Science*, 314, 1286–9

p. 70   **'censorship':** e.g. Revkin, A., 2006: 'Climate expert says NASA tried to silence him', *New York Times*, 29 January 2006

p. 70   **'perilously close':** Hansen, J., et al., 2007: 'Climate change and trace gases', *Philosophical Transactions of the Royal Society A*, 365, 1925–54

p. 71   **'trigger mechanism':** Ibid.

p. 71   **5 metres:** Hansen, J., 2007: 'Scientific reticence and sea level rise', *Environmental Research Letters*, April–June 2007, accessible at http://www.iop. org/EJ/article/1748-9326/2/2/024002/erl7_2_024002.html

pp. 71–2 **3.3 mm a year:** Rahmstorf, S., et al., 2007: 'Recent climate observations compared to projections', *Science*, 316, 709

p. 73   **new Arctic rush:** Krauss, C., et al., 2005: 'As polar ice turns to water, dreams of treasure abound', *New York Times*, 10 October 2005

p. 74 **"'claiming this territory'":** Solovyov, D., 2007: 'Russia explorers snub critics in North Pole row', Reuters, 8 August 2007

p. 74 **Hudson Bay polar bears:** Stirling, I., et al., 1999: 'Long-term trends in the population ecology of polar bears in western Hudson Bay in relation to climate change', *Arctic*, 52, 294–306

p. 75 **ice-free water:** Comiso, J., 2005: 'Impact studies of a 2°C global warming on the Arctic sea ice cover', in *2° Is Too Much! Evidence and Implications of Dangerous Climate Change in the Arctic*, WWF International Arctic Programme, January 2005

p. 75 **desperate walruses:** Krauss, C., et al., 2005: 'Old ways of life are fading as the Arctic thaws', *New York Times*, 20 October 2005

p. 75 **On land:** Arctic Climate Impact Assessment, 2004: *Impacts of a Warming Arctic*, Cambridge University Press – chapter 7: Arctic Tundra and Polar Desert Ecosystems

p. 75 **Freshwater fish:** Ibid. – chapter 8: Freshwater Ecosystems and Fisheries

p. 76 **tundra almost completely disappears:** Kaplan, J., 2005: 'Climate change and Arctic vegetation', in *2° Is Too Much! Evidence and Implications of Dangerous Climate Change in the Arctic*, WWF International Arctic Programme, January 2005

p. 76 **permafrost boundary retreats:** Arctic Climate Impact Assessment, 2004: *Impacts of a Warming Arctic*, Cambridge University Press – chapter 6: Cryosphere and Hydrology

p. 76 **'Arctic amplifier':** New, M., 2005: 'Arctic climate change with a 2°C global warming', in *2° Is Too Much! Evidence and Implications of Dangerous Climate Change in the Arctic*, WWF International Arctic Programme, January 2005

p. 76 **short of terms:** 2005: 'Inuit translators, elders meet to develop Inuktitut words for climate change', Canadian Press, 4 October 2005

p. 76 *uggianaqtuq:* Cowen, R., 2005: 'In melting Arctic, warming is now', *Christian Science Monitor*, 18 October 2005

p. 77 **'Arctic indigenous peoples':** Doyle, A., 2005: 'Arctic peoples urge UN aid to protect cultures', Reuters, 7 December 2005

p. 77 **Rajendra Pachauri:** 'India unlikely to agree emissions caps post Kyoto', Webindia123, 24 November 2005, http://news.webindia123.com/news/showdetails.asp?id=17 1704&n_date=20051124&cat=India

p. 78 **India overtook Japan:** 2005: 'India takes a hard line on global warming', Reuters, 19 November 2005

p. 78 **'no way':** Ibid.

p. 78 **decreases in agricultural output:** undated: *Climate Change Impacts on Agriculture in India*, Keysheet 6, Defra

p. 78 **Forest types:** undated: *Climate Change Impacts on Forestry in India*, Keysheet 7, Defra

p. 78 **major agricultural impacts:** Kavi Kumar, K., and Parikh, J., 2001: 'Indian agriculture and climate sensitivity', *Global Environmental Change*, 11, 147–54

p. 79 **strengthening of the monsoon:** May, W., 2004: 'Simulation of the variability and extremes of daily rainfall during the Indian summer monsoon for present and future times in a global time-slice experiment', *Climate Dynamics*, 22, 183–204

p. 79 **Bangladesh:** Agrawala, S., et al., 2003: 'Development and climate change in Bangladesh: Focus on coastal flooding and the Sundarbans', OECD Environment Directorate, 70pp

p. 79 **having an effect:** Goswami, B., et al., 2006: 'Increasing trend of extreme rain events over India in a warming environment', *Science*, 314, 1442–5

p. 80 **Nepal:** Agrawala, S., et al., 2003: 'Development and climate change in Nepal: Focus on water resources and hydropower', OECD Environment Directorate, 64pp

p. 80 **Andean ice fields:** Barnett, T., et al., 2005: 'Potential impacts of a warming climate on water availability in snow-dominated regions', *Nature*, 438, 303–9

p. 81 **Rio Santa:** Kaser, G., et al., 2003: 'The impact of glaciers on the runoff and the reconstruction of mass balance history from hydrological data in the tropical Cordillera Blanca, Peru', *Journal of Hydrology*, 282, 1, 130–44

p. 82 **channelled … fields:** Chevallier, P., et al., 2004: 'Climate change impact on the water resources from the mountains in Peru', paper presented to the OECD Global Forum on Sustainable Development: Development and Climate Change, Paris, 11–12 November 2004

p. 82 **fall in glacial run-off:** Juen, I., Kaser, G., and Georges, C., 2006: 'Modelling observed and future runoff from a glacierized tropical catchment (Cordillera Blanca, Peru)', *Global and Planetary Change*, 59, 1–4, 37–48

p. 82 **Moche … and Chimu:** Dillehay, T., et al., 2004: 'Pre-industrial human and environment interactions in northern Peru during the late Holocene', *The Holocene*, 14, 2, 272–81

p. 83 **linked to droughts:** Ibid.

p. 84 **already rising:** Bradley, R., et al., 2006: 'Threats to water supplies in the tropical Andes', *Science*, 312, 1755–6

p. 84 **retreating fast:** Ibid.

p. 85 **stored in snowpack:** Mote, P., et al., 2005: 'Declining mountain snowpack in western North America', *Bulletin of the American Meteorological Society*, January 2005

p. 86 **'lizards and tumbleweed':** Barbassa, J., 2005: 'An uncertain future for San Joaquin River', Associated Press, 17 September 2005

p. 87 **projected snowpack declines:** Hayhoe, K., et al., 2004: 'Emissions pathways, climate change, and impacts on California', *Proceedings of the National Academy of Sciences*, 101, 34, 12422–7

p. 87 **Oregon and Washington:** Ruby Leung, L., et al., 2004: 'Mid-century ensemble regional climate change scenarios for the western United States', *Climatic Change*, 62, 75–113

p. 87 **mountainous regions:** Kim, J., 2005: 'A projection of the effects of the climate change induced by increased CO_2 on extreme hydrologic events in the western US', *Climatic Change*, 68, 153–68

p. 87 **Columbia River system:** Ruby Leung, L., et al., 2004: 'Mid-century ensemble regional climate change scenarios for the western United States', *Climatic Change*, 62, 75–113

p. 89 **expected to double:** Southworth, J., et al., 2002: 'Sensitivity of winter wheat yields in the Midwestern United States to future changes in climate, climate variability and CO_2 fertilisation', *Climate Research*, 22, 73–86

p. 89 **Citrus growers:** Tubiello, F., et al., 2002: 'Effects of climate change on US crop production: simulation results using two different GCM scenarios. Part 1: wheat, potato, maize and citrus', *Climate Research*, 20, 259–70

p. 89 **the UK:** undated: *Climate Change and Agriculture in the United Kingdom*, MAFF, 65pp

p. 89 **across Europe:** Maracchi, G., et al., 2005: 'Impacts of present and future climate variability on agriculture and forestry in the temperate regions: Europe', *Climatic Change*, 70, 117–35

p. 89 **Central and South America:** Jones, P., and Thornton, P., 2003: 'The potential impacts of climate change on maize production in Africa and Latin America in 2055', *Global Environmental Change*, 13, 51–9

p. 90 **Mali:** Butt, T., et al., 2005: 'The economic and food security implications of climate change in Mali', *Climatic Change*, 68, 355–78

p. 90 **Botswana:** Chipanshi, A., et al., 2003: 'Vulnerability assessment of the maize and sorghum crops to climate change in Botswana', *Climatic Change*, 61, 339–60

p. 90 **Congo:** Wilkie, D., et al., 1999: 'Wetter isn't better: global warming and food security in the Congo basin', *Global Environmental Change*, 9, 323–8

p. 90 **south-eastern America:** Carbone, G., et al., 2003: 'Response of soybean and sorghum to varying spatial scales of climate change scenarios in the southeastern United States', *Climatic Change*, 60, 73–98

p. 90 **Canada:** 2005: *Implications of a 2°C Global Temperature Rise for Canada's Natural Resources*, WWF, 30 November 2005, 109pp

p. 90 **North Sea cod:** Clark, R., et al., 2003: 'North Sea cod and climate change – modelling the effects of temperature on population dynamics', *Global Change Biology*, 9, 1669–80

p. 91 **South Africa:** Midgeley, G., et al., 2002: 'Assessing the vulnerability of species richness to anthropogenic climate change in a biodiversity hotspot', *Global Ecology & Biogeography*, 11, 445–51

p. 91 **Queensland rainforest:** Williams, S., et al., 2003: 'Climate change in Australian tropical rainforests: an impending environmental catastrophe', *Proceedings of the Royal Society of London B*, 270, 1887–92

p. 92 **Monteverde harlequin frog:** Pounds, A., and Puschendorf, R., 2004: 'Clouded futures', *Nature*, 427, 107–9

p. 92 **leaves them stranded:** Hare, W., 2003: *Assessment of Knowledge on Impacts of Climate Change – Contribution to* *the Specification of Art. 2 of the UNFCCC*, WBGU, Berlin, 104pp

p. 92 **100–1,000 times greater:** Martens, P., et al., 2003: 'Biodiversity: luxury or necessity?', *Global Environmental Change*, 13, 75–81

p. 92 **towards the poles:** Parmesan, C., and Yohe, P., 2003: 'A globally coherent fingerprint of climate change impacts across natural systems', *Nature*, 421, 37–42

p. 92 **range and behaviour:** Root, T., et al., 2003: 'Fingerprints of global warming on wild animals and plants', *Nature*, 421, 57–60

p. 93 **English blue butterfly:** Thomas, J., 2006: 'Biodiversity and climate change', lecture at Oxford University, 23 January 2006

p. 93 **30 kilometres per decade:** Calculation as follows: climate zones shift 150 km per degree C. So this is about 300 km for 2°C warming, relevant to this chapter, or 30 km per decade, assuming 2°C only by 2100 and no additional UK temperature increases over global.

p. 94 **pied flycatchers:** Both, C., et al., 2006: 'Climate change and population declines in a long-distance migratory bird', *Nature*, 441, 81–3

p. 95 **over a third of all species:** Thomas, C., et al., 2004: 'Extinction risk from climate change', *Nature*, 427, 145–8

p. 95 **'over a million species':** 2004: 'Climate change threatens a million species with extinction', University of Leeds press release, 7 January 2004

3 THREE DEGREES

p. 103 **split in half:** Hoerling, M., et al., 2005: 'Detection and attribution of 20th century northern and southern

African monsoon change', *Journal of Climate*, 19, 16, 3989–4008

p. 104 **total remobilisation:** Thomas, D., et al., 2005: 'Remobilization of southern African desert dune systems by twenty-first century global warming', *Nature*, 435, 1218–21

p. 108 **evolved bipedalism:** Haywood, A., and Valdes, P., 2006: 'Vegetation cover in a warmer world simulated using a dynamic global vegetation model for the mid-Pliocene', *Palaeogeography, Palaeoclimatology, Palaeoecology*, 237, 412–77

p. 108 **fossil wood and leaves:** Francis, J., and Hill, R., 1996: 'Fossil plants from the Pliocene Sirius Group, Transantarctic Mountains; evidence for climate from growth rings and fossil leaves', PALAIOS, 11, 4, 389–96

p. 109 **variety of mammals:** Tedford, R., and Harington, R., 2003: 'An Arctic mammal fauna from the early Pliocene of North America', *Nature*, 425, 388–90

p. 109 **25 metres higher:** Barreiro, M., et al., 2006: 'Simulation of warm tropical conditions with application to middle Pliocene atmospheres', *Climate Dynamics*, 26, 249–365

p. 110 **reconstruction of sea temperatures:** Haywood, A., et al., 2005: 'Warmer tropics during the mid-Pliocene? Evidence from alkenone paleothermometry and a fully coupled ocean-atmosphere GCM', *Geochemistry, Geophysics, Geosystems*, 6, 3, 20pp

p. 110 **'critical to understanding':** British Antarctic Survey, 2005: 'Carbon dioxide role in past climate revealed', press release, 11 April 2005

p. 110 **seasonally ice-free:** Haywood, A., et al., 2005: 'Warmer tropics during the mid-Pliocene? Evidence from alkenone paleothermometry and a fully coupled ocean-atmosphere

GCM', *Geochemistry, Geophysics, Geosystems*, 6, 3, 20pp

p. 110 **Atlantic circulation ... reduced:** Raymo, M., et al., 1996: 'Mid-Pliocene warmth: stronger greenhouse and stronger conveyor', *Marine Micropaleontology*, 27, 313–26

p. 111 **360 to 400 ... ppm:** Haywood, A., and Williams, M., 2005: 'The climate of the future: clues from three million years ago', *Geology Today*, 21, 4, 138–43

p. 111 **just under three degrees:** Haywood, A., and Valdes, P., 2004: 'Modelling Pliocene warmth: contribution of atmosphere, oceans and cryosphere', *Earth and Planetary Science Letters*, 218, 363–77; Jiang, D., et al., 2005: 'Modeling the middle Pliocene climate with a global atmospheric general circulation model', *Journal of Geophysical Research*, 110, D14107

p. 112 **storms ... famine:** Couper-Johnston, R., 2000: *El Niño: The Weather Phenomenon that Changed the World*, Hodder & Stoughton

p. 113 **drove icebergs:** Ibid.

p. 113 **always been accompanied:** Krishna Kumar, K., et al., 2006: 'Unraveling the mystery of Indian monsoon failure during El Niño', *Science*, 314, 115–19

p. 114 **near-permanent El Niño:** Boer, G., et al., 2004: 'Is there observational support for an El Niño-like pattern of future global warming?', *Geophysical Research Letters*, 31, L06201

p. 114 **weaker El Niño:** Meehl, G., et al., 2006: 'Future changes of El Niño in two global coupled climate models', *Climate Dynamics*, 26, 549–66

p. 114 **little change:** Collins, M., and the CMIP Modelling Groups, 2005: 'El Niño or La Niña-like climate change', *Climate Dynamics*, 24, 89–104; Cane, M., 2005: 'The evolution of El Niño, past and future', *Earth and*

Planetary Science Letters, 230, 227–40

p. 114 **weaker or absent:** Boer, G., et al., 2004: 'Is there observational support for an El Niño-like pattern of future global warming?', *Geophysical Research Letters*, 31, L06201

p. 114 **winds weakened or collapsed:** Barreiro, M., et al., 2006: 'Simulation of warm tropical conditions with application to middle Pliocene atmospheres', *Climate Dynamics*, 26, 249–365

p. 115 **winds ... begun to slacken:** Vecchi, G., et al., 2006: 'Weakening of tropical Pacific atmospheric circulation due to anthropogenic forcing', *Nature*, 441, 73–6

p. 115 **'super El Niños':** Hansen, J., et al., 2006: 'Global temperature change', *Proceedings of the National Academy of Sciences*, 103, 39, 14288–93

p. 115 **Europe ... drier winters:** Müller, W., and Roeckner, E., 2006: 'ENSO impact on midlatitude circulation patterns in future climate change projections', *Geophysical Research Letters*, 33, L05711

p. 115 **wind shear:** Vecchi, G., and Soden, B., 2007: 'Increased tropical Atlantic wind shear in model projections of global warming', *Geophysical Research Letters*, 34, L08702

p. 115 **dry areas of California:** Maurer, E., et al., 2006: 'Amplification of streamflow impacts of El Niño by increased atmospheric greenhouse gases', *Geophysical Research Letters*, 33, L02707

p. 118 **1.5°C boost:** Cox, P., et al., 2000: 'Acceleration of global warming due to carbon cycle feedbacks in a coupled climate model', *Nature*, 408, 184–7

p. 118 **carbon stored:** Fung, I., et al., 2005: 'Evolution of carbon sinks in a changing climate', *Proceedings of the National Academy of Sciences*, 102, 32, 11201–6

p. 118 **French team repeated:** Berthelot, M., et al., 2005: 'How uncertainties in future climate change predictions translate into future terrestrial carbon fluxes', *Global Change Biology*, 11, 1–12

p. 118 **Another team:** Zeng, N., et al., 2004: 'How strong is carbon cycle feedback under global warming?', *Geophysical Research Letters*, 31, L20203

p. 119 **models investigated:** Li, W., et al., 2007: 'Future precipitation changes and their implications for tropical peatlands', *Geophysical Research Letters*, 34, L01403

p. 119 **'hydrological power engine':** Bunyard, P., 2002: 'Climate and the Amazon: Consequences for our planet', *The Ecologist*, October 2002

p. 120 **Brazilian government:** Ibid.

p. 120 **rainfall declines:** Betts, R., et al., 2004: 'The role of ecosystem-atmosphere interactions in simulated Amazonian precipitation decrease and forest dieback under global climate warming', *Theoretical and Applied Climatology*, 78, 137–56

p. 120 **essentially desert:** Cowling, S., et al., 2004: 'Contrasting simulated past and future responses of the Amazonian forest to atmospheric change', *Philosophical Transactions of the Royal Society of London B*, 359, 539–47

p. 120 **Every fire season:** Barlow, J., and Peres, C., 2004: 'Ecological responses to El Niño-induced surface fires in central Brazilian Amazonia: management implications for flammable tropical forests', *Philosophical Transactions of the Royal Society of London B*, 359, 367–80

p. 122 **New South Wales:** Hennessy, K., et al., 2004: *Climate Change in New South Wales: Part 2 – Projected changes in climate extremes*, CSIRO, November 2004, 79pp

p. 123 **drought frequency could triple:** Ibid.

p. 123 **rainfall ... could plummet:** Hennessy, K., et al., 2004: *Climate Change in New South Wales: Part 1 – Past climate variability and projected changes in average climate*, CSIRO, November 2004, 46pp; see p. 36 for projection relevant to this chapter, using 550 stabilisation high scenario at 2070.

p. 123 **Victoria:** Suppiah, R., et al., 2004: *Climate Change in Victoria: Assessment of climate change for Victoria: 2001–2002*, CSIRO, April 2004, 33pp

p. 123 **South-western Australia:** Timbal, B., 2004: 'Southwest Australia past and future rainfall trends', *Climate Research*, 26, 233–49

p. 123 **elderly people will die:** Woodruff, R., et al., 2005: *Climate Change Health Impacts in Australia: Effects of Dramatic CO_2 Emissions Reductions*, Australian Conservation Foundation and Australian Medical Association, 45pp

p. 123 **dengue fever:** Ibid.

p. 123 **350 litres of water:** Australian Natural Resources Atlas – Water, Australian Government Department of the Environment and Heritage, http://audit.deh.gov.au/ANRA/water/water_frame.cfm?region_type=AUS&region_code=AUS&info=allocation

p. 123 **Darwin and Queensland:** Pittock, B. (ed.), 2003: *Climate Change – An Australian Guide to the Science and Potential Impacts*, Australian Greenhouse Office, p. 88

p. 123 **remain viable:** Luo, Q., et al., 2007: 'Risk analysis of possible impacts of climate change on South Australian wheat production', *Climatic Change*, 85, 1–2, 89–101

p. 124 **Murray-Darling River basin:** Pittock, B. (ed.), 2003: *Climate Change – An Australian Guide to the Science and Potential Impacts*, Australian Greenhouse Office, p.88, based on A1 mid scenario by 2100

p. 124 **drop in annual rainfall:** Hope, P., 2006: 'Projected future changes in synoptic systems influencing southwest Western Australia', *Climate Dynamics*, 26, 765–80

p. 125 **Mount Pinatubo:** Fromm, M., et al., 2006: 'Violent pyro-convective storm devastates Australia's capital and pollutes the stratosphere', *Geophysical Research Letters*, 33, L05815

p. 128 **increased hurricane intensity:** Knutson, T., and Tuleya, R., 2004: 'Impact of CO_2-induced warming on simulated hurricane intensity and precipitation: sensitivity to the choice of climate model and convective parameterisation', *Journal of Climate*, 17, 18, 3477–95

p. 129 **Hurricanes:** Emanuel, K., 2005: *Divine Wind: The History and Science of Hurricanes*, Oxford University Press, p. 258

p. 129 **exacerbate the drying:** Sewall, J., and Sloan, L., 2004: 'Disappearing Arctic sea ice reduces available water in the American west', *Geophysical Research Letters*, 31, L06209

p. 129 **80 per cent:** Johannessen, O., et al., 2004: 'Arctic climate change: observed and modelled temperature and sea ice variability', *Tellus*, 26A, 328–41

p. 130 **'new Arctic state':** Stroeve, J., et al., 2007: 'Arctic sea ice decline: Faster than forecast', *Geophysical Research Letters*, 34, L09501

p. 130 **Smaller ice caps:** Oerlemans, J., et al., 2005: 'Estimating the contribution of Arctic glaciers to sea-level change in the next 100 years', *Annals of Glaciology*, 42, 1, 230–6

p. 130 **3,500 cubic kilometres:** CWE Glaciers Group, 2004: *The Impact of Climate Change on Glaciers in the Nordic Countries*, The CWE Project, 42pp

p. 131 **rise by a quarter:** Raper, S., and Braithwaite, R., 2005: 'The potential for sea level rise: New estimates from glacier and ice cap area and volume distributions',

Geophysical Research Letters, 32, L05502

p. 131 **Norway enjoys a growing season:** Skaugen, T., and Tveito, O., 2004: 'Growing-season and degree-day scenario in Norway for 2021–2050', *Climate Research*, 26, 221–32

p. 131 **Finland:** Dankers, R., and Christensen, O., 2005: 'Climate change impact on snow coverage, evaporation and river discharge in the sub-Arctic Tanna basin, Northern Fennoscandia', *Climatic Change*, 69, 367–92

p. 133 **revolt or even regicide:** Diamond, J., 2005: *Collapse: How Societies Choose to Fail or Survive*, Allen Lane, chapter 5 – The Maya Collapses

p. 133 **Lake-floor records:** Hodell, D., et al., 2005: 'Terminal Classic drought in the northern Maya lowlands inferred from multiple sediment cores in Lake Chichancanab (Mexico)', *Quaternary Science Reviews*, 25, 12–13, 1413–27

p. 133 **'intense multiyear droughts':** Huag, G., et al., 2003: 'Climate and the collapse of Maya civilization', *Science*, 299, 1731–5

p. 134 **1–2 mm per day:** Johns, T., et al., 2003: 'Anthropogenic climate change for 1860–2100 simulated with the HadCM3 model under updated emissions scenarios', *Climate Dynamics*, 20, 583–612 – see graph figure 13-c for the B1 world.

p. 134 **worsen deforestation:** Breshears, D., et al., 2005: 'Regional vegetation die-off in response to global-change-type drought', *PNAS*, 102, 42, 15144–8

p. 134 **climate change 'hot spots':** Giorgi, F., 2006: 'Climate change hot spots', *Geophysical Research Letters*, 33, L08707

p. 135 **times in the past:** Burns, S., et al., 2003: 'Indian Ocean climate and absolute chronology over Dansgaard/Oeschger Events 9 to 13', *Science*, 301, 1365–7

p. 136 **'Asian Brown Cloud':** Zickfield, K., et al., 2005: 'Is the Indian summer monsoon stable against global change?', *Geophysical Research Letters*, 32, L15707

p. 136 **many different studies:** May, W., 2004: 'Simulation of the variability and extremes of daily rainfall during the Indian summer monsoon for present and future times in a global time-slice experiment', *Climate Dynamics*, 22, 183–204; Ueda, H., et al., 2006: 'Impact of anthropogenic forcing on the Asian summer monsoon as simulated by 8 GCMs', *Geophysical Research Letters*, 33, L06703

p. 136 **associated extreme flooding:** Dairaku, K., and Emori, S., 2006: 'Dynamic and thermodynamic influences on intensified daily rainfall during the Asian summer monsoon under doubled atmospheric CO_2 conditions', *Geophysical Research Letters*, 33, L01704

p. 136 **even more difficult:** Meehl, G., and Arblaster, J., 2003: 'Mechanisms for projected future changes in south Asian monsoon precipitation', *Climate Dynamics*, 21, 659–75

p. 136 **harsh drought:** Abram, N., et al., 2007: 'Seasonal characteristics of the Indian Ocean Dipole during the Holocene epoch', *Nature*, 445, 299–302; Overpeck, J., and Cole, J., 2007: 'Lessons from a distant monsoon', *Nature*, 445, 270–1

p. 139 **accelerated glacial retreat:** WWF Nepal Program, 2005: *An Overview of Glaciers, Glacier Retreat, and Subsequent Impacts in Nepal, India and China*, WWF, March 2005, 70pp

p. 139 **decreasing water supplies:** Rees, G., and Collins, D., 2004: *An Assessment of the Potential Impacts of Deglaciation on the Water Resources of the Himalaya*, DFID KAR Project No. R7980, 54pp and Annexes

p. 140 **China:** Barnett, T., et al., 2005: 'Potential impacts of a warming climate on water availability in snow-dominated regions', *Nature*, 438, 303–9

p. 142 **Colorado River:** Christensen, N., et al., 2004: 'The effect of climate change on the hydrology and water resources of the Colorado river basin', *Climatic Change*, 62, 337–63

p. 142 **Lake Mead:** Allen, J., 2003: 'Drought lowers Lake Mead', NASA Earth Observatory, http://earthobservatory.nasa.gov/Study/LakeMead/

p. 142 **system would essentially fail:** Christensen, N., et al., 2004: 'The effect of climate change on the hydrology and water resources of the Colorado river basin', *Climatic Change*, 62, 337–63

p. 142 **Glen Canyon:** For more see the Glen Canyon Institute website: http://www.glencanyon.org/

p. 143 **date of spring snowmelt:** Stewart, I., et al., 2004: 'Changes in snowmelt runoff timing in western North America under a "business as usual" climate change scenario', *Climatic Change*, 62, 217–32

p. 143 **passed on to the oceans:** Barnett, T., et al., 2005: 'Potential impacts of a warming climate on water availability in snow-dominated regions', *Nature*, 438, 303–9

p. 143 **Pacific coast:** Mote, P., et al., 2003: 'Preparing for climatic change: The water, salmon and forests of the Pacific northwest', *Climatic Change*, 61, 45–88

p. 144 **Canadian river:** Lapp, S., et al., 2005: 'Climate warming impacts on snowpack accumulation in an alpine watershed', *International Journal of Climatology*, 25, 521–36

p. 144 **Rockies:** Laprise, R., et al., 2003: 'Current and perturbed climate as simulated by the second-generation Canadian Regional Climate Model (CRCM-II) over northwestern North America', *Climate Dynamics*, 21, 405–21

p. 144 **Dust Bowl years:** Seager, R., et al., 2007: 'Model projections of an imminent transition to a more arid climate in Southwestern North America', *Science*, 316, 1181–4

p. 144 **severe wildfires:** Westerling, A., et al., 2006: 'Warming and earlier spring increase western US forest wildfire activity', *Science*, 313, 940–3

p. 144 **out-of-control wildfires:** Fried, J., et al., 2004: 'The impact of climate change on wildfire severity: a regional forecast for northern California', *Climatic Change*, 64, 169–91

p. 144 **more of high fire danger:** Brown, T., et al., 2004: 'The impact of twenty-first century climate change on wildland fire danger in the western United States: An applications perspective', *Climatic Change*, 62, 365–88

p. 146 **half a metre higher:** Gornitz, V., et al., 2002: 'Impacts of sea level rise in the New York City metropolitan area', *Global and Planetary Change*, 32, 61–88

p. 147 **one-in-100-year flood:** Ibid.

p. 147 **beaches … moving inland:** Ibid.

p. 147 **flood barriers:** Van Lenten, C., 2005: 'Storm surge barriers for New York harbour?', *Science and the City*, http://www.nyas.org/snc/update.asp?UpdateID=28

p. 147 **Malcolm Bowman:** First quote in above, second in Bowman, M., 2005: 'A city at sea', *New York Times*, 25 September 2005.

p. 148 **200 times:** 2002: *London's Warming: The Impacts of Climate Change on London*, The London Climate Change Partnership, October 2002, 293pp

p. 148 **flooding events:** Lowe, J., et al., 2001: 'Changes in occurrence of storm surges around the United Kingdom under a future climate scenario using a dynamic storm surge model driven by the Hadley Centre models', *Climate Dynamics*, 18, 179–88

p. 149 **more storm wind events:** Leckebusch, G., and Ulbrich, U., 2004: 'On the relationship between cyclones and extreme windstorm events over Europe under climate change', *Global and Planetary Change*, 44, 181–93

p. 149 **more frequent worldwide:** Lambert, S., and Fyfe, J., 2006: 'Changes in winter cyclone frequencies and strengths simulated in enhanced greenhouse warming experiments: results from the models participating in the IPCC diagnostic exercise', *Climate Dynamics*, 26, 713–28

p. 149 **stronger storm surges:** Beniston, M., et al., 2004: 'Future extreme events in a European climate: An exploration of regional climate model projections', Prudence WP5

p. 149 **intensification of the hydrological cycle:** Huntingdon, T., 2005: 'Evidence for intensification of the global water cycle: Review and synthesis', *Journal of Hydrology*, 319, 83–95

p. 150 **increase in precipitation:** Zhang, X., et al., 2007: 'Detection of human influence on twentieth-century precipitation trends', *Nature*, 448, 461–5

p. 150 **increasingly intense heavier events:** Semmler, T., and Jacob, D., 2004: 'Modeling extreme precipitation events – a climate change simulation for Europe', *Global and Planetary Change*, 44, 119–27

p. 150 **increase in peak flooding:** Kay, A., et al., 2006: 'RCM rainfall for UK flood frequency estimation. II. Climate change results', *Journal of Hydrology*, 318, 163–72

p. 150 **deluges and water shortages:** Pal, J., et al., 2004: 'Consistency of recent European summer precipitation trends and extremes with future regional climate projections', *Geophysical Research Letters*, 31, L13202

p. 150   **Alps:** Jasper, K., et al., 2004: 'Differential impacts of
         climate change on the hydrology of two alpine river
         basins', *Climate Research*, 26, 113–29

p. 150   **Rhine:** Shabalova, M., et al., 2003: 'Assessing future
         discharge of the river Rhine using regional climate model
         integrations and a hydrological model', *Climate Research*,
         23, 233–46

p. 150   **Mediterranean:** Gibelin, A.-L., and Déqué, M., 2003:
         'Anthropogenic climate change over the Mediterranean
         region simulated by a global variable resolution model',
         Climate Dynamics, 20, 237–339

p. 151   **rainfall and frequent floods:** McHugh, M., 2005:
         'Multi-model trends in East African rainfall associated
         with increased CO_2', *Geophysical Research Letters*, 32,
         L01707

p. 151   **85 per cent of infections:** 2002: *The World Health Report
         2002: Reducing Risks, Promoting Healthy Life*, World
         Health Organization

p. 152   **96 per cent of the country:** Ebi, K., et al., 2005: 'Climate
         suitability for stable malaria transmission in Zimbabwe
         under different climate change scenarios', *Climatic
         Change*, 73, 375–93

p. 152   **hundreds of millions more people:** van Leishout, M., et
         al., 2004: 'Climate change and malaria: analysis of the
         SRES climate and socio-economic scenarios', *Global
         Environmental Change*, 14, 87–99

p. 152   **more people exposed:** Ibid.

p. 154   **tepui endemic plants:** Rull, V., and Vegas-Vilarrubia, T.,
         2006: 'Unexpected biodiversity loss under global warm-
         ing in the neotropical Guyana highlands: a preliminary
         appraisal', *Global Change Biology*, 12, 1–9

p. 154   **coral reefs:** Donner, S., et al., 2005: 'Global assessment of coral bleaching and required rates of adaptation under climate change', *Global Change Biology*, 11, 2251–65

p. 155 **Europe's plants:** Thuiller, W., et al., 2005: 'Climate change threats to plant diversity in Europe', *PNAS*, 102, 23, 8245–50

p. 155 **Rockies and Great Plains:** Townsend Peterson, A., 2003: 'Projected climate change effects on Rocky Mountain and Great Plains birds: generalities of biodiversity consequences', *Global Change Biology*, 9, 647–55

p. 155 **north-eastern China:** Shao, G., et al., 2003: 'Sensitivities of species compositions of the mixed forest in eastern Eurasian continent to climate change', *Global and Planetary Change*, 37, 307–13

p. 156 **'novel' climates:** Williams, J., Jackson, S., and Kutzbach, J., 2007: 'Projected distributions of novel and disappearing climates by 2100 AD', *PNAS*, 104, 14, 5738–42

p. 156 **between a third and a half:** Thomas, C., et al., 2004: 'Extinction risk from climate change', *Nature*, 427, 145–8

p. 157 **'decline by 10 per cent':** Halweil, B., 2005: 'The irony of climate', *Worldwatch Magazine*, March/April 2005

p. 157 **Africa's semi-arid tropics:** Sivakumar, M., et al., 2005: 'Impacts of present and future climate variability and change on agriculture and forestry in the arid and semi-arid tropics', *Climatic Change*, 70, 31–72

p. 157 **crippling declines:** Easterling, W., and Apps, M., 2005: 'Assessing the consequences of climate change for food and forest resources: a view from the IPCC', *Climatic Change*, 70, 165–89

p. 158 **'major planetary granaries':** Ibid.

p. 158 **United States, southern areas:** Tsvetsinskaya, A., et al., 2003: 'The effect of spatial scale of climate change scenarios on simulated maize, winter wheat and rice production in the southeastern United States', *Climatic Change*, 60, 37–71

p. 158 **Additional flood damages:** Rosenzweig, C., et al., 2002: 'Increased crop damage in the US from excess precipitation under climate change', *Global Environmental Change*, 12, 197–202

p. 158 **'shift hundreds of kilometres':** Motha, R., and Baier, W., 2005: 'Impacts of present and future climate change and climate variability on agriculture in the temperate regions: North America', *Climatic Change*, 70, 137–64

p. 158 **drive up market prices:** Easterling, W., and Apps, M., 2005: 'Assessing the consequences of climate change for food and forest resources: a view from the IPCC', *Climatic Change*, 70, 165–89

4 FOUR DEGREES

p. 164 **50 centimetres:** El Raey, M., et al., 1999: 'Adaptation to the impacts of sea level rise in Egypt', *Climate Research*, 12, 117–28

p. 165 **Boston:** Kirshen, P., et al., undated: *Climate's Long-Term Impacts on Metro Boston (CLIMB)*, media summary

p. 165 **New Jersey:** Cooper, M., et al., 2005: *Future Sea Level Rise and the New Jersey Coast: Assessing Potential Impacts and Opportunities*, Science, Technology and Environmental Policy Program, Woodrow Wilson School of Public and International Affairs, Princeton University, June 2005

p. 167 **West Antarctic Ice Sheet:** Mercer, J. H., 1978: 'West Antarctic ice sheet and CO_2 greenhouse effect: A threat of disaster', *Nature*, 271, 321–5

p. 168 **speed up and retreat:** Payne, A., et al., 2004: 'Recent dramatic thinning of the largest West Antarctic ice stream triggered by oceans', *Geophysical Research Letters*, 31, L23401

p. 168 **losing ... ice:** Velicogna, I., and Wahr, J., 2006: 'Measurements of time-variable gravity show mass loss in Antarctica', *Science*, 311, 1754–6

p. 168 **propagated far inland:** Kerr, R., 2004: 'A bit of icy Antarctica is sliding towards the sea', *Science*, 305, 1897

p. 168 **adding 0.14 millimetres:** Thomas, R., et al., 2004: 'Accelerated sea-level rise from West Antarctica', *Science*, 306, 255–8; van den Broeke, M., et al., 2006: 'Snowfall in West Antarctica much greater than previously assumed', *Geophysical Research Letters*, 33, L02505. The latter gives a newer update.

p. 169 **'monitoring':** NASA mission news, 15 May 2007: 'NASA finds vast regions of West Antarctica melted in recent past', available at http://www.nasa.gov/vision/earth/lookingatearth/arctic-20070515.html

p. 169 **accelerated their flow:** Alley, R., et al., 2005: 'Ice-sheet and sea-level changes', *Science*, 310, 456–60

p. 169 **Ross and the Ronne:** Oppenheimer, M., and Alley, R., 2005: 'Ice sheets, global warming, and Article 2 of the UNFCCC', *Climatic Change*, 68, 257–67

p. 170 **submarine beds:** Bindshadler, R., 2006: 'Hitting the ice sheets where it hurts', *Science*, 311, 1720–1

p. 170 **nearly 40 million years:** Zachos, J., et al., 2001: 'Trends, rhythms, and aberrations in global climate over 65 Ma to present', *Science*, 292, 686–93

p. 171 **China's coastal waters:** Liu, J., and Diamond, J., 2005: 'China's environment in a globalizing world', *Nature*, 435, 1179–86

p. 172 **voracious as Americans:** Brown, L., 2005: 'Learning from China: Why the Western economic model will not work for the world', Earth Policy Institute, Eco-Economy Update, 9 March 2005; http://www.earth-policy.org/Updates/2005/Update46.htm

p. 172 **agricultural production will crash:** China–UK collaboration project, 2004: *Investigating the Impacts of Climate Change on Chinese Agriculture*, Defra

p. 173 **Australia:** Hare, B., 2006: 'Relationship between increases in global mean temperature and impacts on ecosystems, food production, water and socio-economic systems', in Schellnhuber, H. J. (ed.), *Avoiding Dangerous Climate Change*, Cambridge University Press

p. 173 **India:** 2005: *Climate Change Scenarios for India*, Keysheet 2, Defra

p. 173 **global analysis:** Wang, G., 2005: 'Agricultural drought in a future climate: results from 15 global climate models participating in the IPCC 4th assessment', *Climate Dynamics*, 29, 739–53

p. 174 **China ... civilisations:** Chang Huang, C., et al., 2003: 'Climatic aridity and the relocations of the Zhou culture in the southern Loess Plateau of China', *Climatic Change*, 61, 361–78

p. 174 **Harappan civilisation:** Staubwasser, M., et al., 2003: 'Climate change at the 4.2 ka BP termination of the Indus valley civilization and the Holocene south Asian monsoon variability', *Geophysical Research Letters*, 30, 8, 1425

p. 177 **rainfall declines:** Raisanen, J., et al., 2004: 'European climate in the late twenty-first century: regional simulations with two driving global models and two forcing scenarios', *Climate Dynamics*, 22, 13–31

p. 177 **heatwaves:** Holt, T., and Palutikof, J., 2004: *The Effect of Global Warming on Heat Waves and Cold Spells in the*

Mediterranean, Prudence Deliverable D5A4, Climatic Research Unit

p. 177 **200 to 500 per cent increase:** Diffenbaugh, N., et al., 2007: 'Heat stress intensification in the Mediterranean climate change hotspot', *Geophysical Research Letters*, 34, L11706

p. 177 **Switzerland:** Beniston, M., and Diaz, H., 2004: 'The 2003 heat wave as an example of summers in a greenhouse climate? Observations and climate model simulations for Basel, Switzerland', *Global and Planetary Change*, 44, 73–81

p. 177 **England:** Brabson, B., et al., 2005: 'Soil moisture and predicted spells of extreme temperatures in Britain', *Journal of Geophysical Research*, 110, D05104

p. 178 **temperatures on the continent:** Rowell, D., 2005: 'A scenario of European climate change for the late twenty-first century: seasonal means and interannual variability', *Climate Dynamics*, 25, 837–49

p. 178 **Caspian Sea:** Elguindi, M., and Giorgi, F., 2006: 'Projected changes in the Caspian Sea level for the 21st century based on the latest AOGCM models', *Geophysical Research Letters*, 33, L08706

p. 178 **'hot spots':** Giorgi, F., 2006: 'Climate change hot spots', *Geophysical Research Letters*, 33, L08707

p. 180 **snow will be a rarity:** Snow cover duration from graph Fig. 3 in Beniston, M., et al., 2003: 'Snow pack in the Swiss Alps under changing climatic conditions: an empirical approach for climate impacts studies', *Theoretical and Applied Climatology*, 74, 19–31

p. 180 **3,000 metres:** Beniston, M., et al., 2003: 'Estimates of snow accumulation and volume in the Swiss Alps under changing climatic conditions', *Theoretical and Applied Climatology*, 76, 125–40

p. 180 **glaciers will vanish:** Zemp, M., et al., 2006: 'Alpine glaciers to disappear within decades?', *Geophysical Research Letters*, 33, 13, L13504

p. 180 **Heatwaves:** Beniston, M., 2005: 'Warm winter spells in the Swiss Alps: Strong heat waves in a cold season? A study focusing on climate observations at the Saentis high mountain site', *Geophysical Research Letters*, 32, L01812

p. 182 **over northern Europe:** Fischer-Bruns, I., et al., 2005: 'Modelling the variability of midlatitude storm activity on decadal to century time scales', *Climate Dynamics*, 25, 461–76

p. 182 **Scotland:** Yin, J., 2005: 'A consistent poleward shift in the storm tracks in simulations of 21st century climate', *Geophysical Research Letters*, 32, L18701

p. 182 **costs ... could increase:** Leckebusch, G., et al., 2007: 'Property loss potentials for European midlatitude storms in a changing climate', *Geophysical Research Letters*, 34, L05703

p. 182 **fewer of them:** Lambert, F., and Fyfe, J., 2006: 'Changes in winter cyclone frequencies and strengths simulated in enhanced greenhouse warming experiments: results from the models participating in the IPCC diagnostic exercise', *Climate Dynamics*, 26, 713–28

p. 182 **south and east of England:** UKCIP, 2002: *Climate Change Scenarios for the United Kingdom*, The UKCIP02 Briefing Report, April 2002

p. 183 **expected to quadruple:** undated: *Future Flooding*, executive summary, Foresight programme, UK Office of Science and Technology

p. 183 **entire English coast:** Ibid.

p. 183 **Snowfall:** UKCIP, 2002: *Climate Change Scenarios for the*

United Kingdom, The UKCIP02 Briefing Report, April 2002

p. 185 **PhD thesis:**
http://www.lib.utexas.edu/etd/d/2005/cooked44152/cooke
d44152.pdf; **paper:** Cooke, J., et al., 2003: 'Precise timing
and rate of massive late Quaternary soil denudation',
Geology, 31, 10, 853–6

p. 185 **higher rainfall intensity:** Chen, M., et al., 2005: 'Changes
in precipitation characteristics over North America for
doubled CO_2', *Geophysical Research Letters*, 32, L19716

p. 185 **entire land surface:** Semenov, V., and Bengtsson, L., 2002:
'Secular trends in daily precipitation characteristics:
greenhouse gas simulation with a coupled AOGCM',
Climate Dynamics, 19, 123–40

p. 185 **central and northern Europe:** Frei, C., et al., 2006: 'Future
change of precipitation extremes in Europe:
Intercomparison of scenarios from regional climate
models', *Journal of Geophysical Research*, 111, D06105

p. 185 **Winter cyclones:** Lambert, F., and Fyfe, J., 2006: 'Changes
in winter cyclone frequencies and strengths simulated in
enhanced greenhouse warming experiments: results from
the models participating in the IPCC diagnostic exercise',
Climate Dynamics, 26, 713–28

p. 185 **eastern Australian region:** Walsh, K., et al., 2004:
'Fine-resolution regional climate model simulations of
the impact of climate change on tropical cyclones near
Australia', *Climate Dynamics*, 22, 47–56

p. 186 **Korea:** Boo, K.-O., et al., 2004: 'Response of global
warming on regional climate change over Korea: An
experiment with the MM5 model', *Geophysical Research
Letters*, 31, L21206

p. 186 **summer North Pole:** Teng, H., et al., 2006: 'Twenty-first
century Arctic climate change in the CCSM3 IPCC

scenario simulations', *Climate Dynamics*, 26, 601–16

p. 186 **permafrost ... shrinks:** Lawrence, D., and Slater, A., 2005: 'A projection of severe near-surface permafrost degradation during the 21st century', *Geophysical Research Letters*, 32, L24401

p. 187 **Russian Far East:** Arctic Climate Impact Assessment, 2004: *Impacts of a Warming Arctic*, Cambridge University Press, pp. 87–8

p. 187 **fresh water draining:** Lawrence, D., and Slater, A., 2005: 'A projection of severe near-surface permafrost degradation during the 21st century', *Geophysical Research Letters*, 32, L24401

p. 188 **carbon ... locked up:** Frey, K., and Smith, L., 2005: 'Amplified carbon release from vast West Siberian peatlands by 2100', *Geophysical Research Letters*, 32, L09401

p. 188 **'unplugging the refrigerator':** Walker, G., 2007: 'A world melting from the top down', *Nature*, 446, 718–21

p. 189 **700 per cent increase:** Frey, K., and Smith, L., 2005: 'Amplified carbon release from vast West Siberian peatlands by 2100', Geophysical Research Letters, 32, L09401.

p. 189 **recently defrosted mire:** Christensen, T., et al., 2005: 'Thawing sub-arctic permafrost: Effects on vegetation and methane emissions', *Geophysical Research Letters*, 31, L04501

p. 189 **thaw lakes:** Walter, K. M., et al., 2006: 'Methane bubbling from Siberian thaw lakes as a positive feedback to climate warming', *Nature*, 443, 71–5

p. 189 **already accelerating:** Torre Jorgenson, M., et al., 2006: 'Abrupt increase in permafrost degradation in Arctic Alaska', *Geophysical Research Letters*, 33, L02503

p. 189 **'potential amounts':** Ibid.

p. 189 **overstated:** Delisle, G., 2007: 'Near-surface permafrost degradation: How severe during the 21st century?', *Geophysical Research Letters*, 34, L09503

p. 190 **'doubled':** Walker, G., 2007: 'A world melting from the top down', *Nature*, 446, 718–21

5 FIVE DEGREES

p. 193 **evaporation and precipitation increase:** Manabe, S., et al., 2004: 'Century-scale change in water availability: CO_2-quadrupling experiment', *Climatic Change*, 64, 59–76

p. 193 **tropics … UK:** Huntingford, C., et al., 2003: 'Regional climate-model predictions of extreme rainfall for a changing climate', *Quarterly Journal of the Royal Meteorological Society*, 129, 1607–21

p. 194 **'expansion of major deserts':** Manabe, S., et al., 2004: 'Century-scale change in water availability: CO_2-quadrupling experiment', *Climatic Change*, 64, 59–76

p. 195 **California's … snowpack:** Hayhoe, K., et al., 2004: 'Emissions pathways, climate change, and impacts on California', *Proceedings of the National Academy of Sciences*, 101, 34, 12422–7

p. 195 **Yemen:** Brown, L., 2003: *Plan B: Rescuing a Planet under Stress and a Civilization in Trouble*, W.W. Norton & Co., pp. 31–2

p. 196 **agricultural production declines:** Parry, M., et al., 2004: 'Effects of climate change on global food production under SRES emissions and socio-economic scenarios', *Global Environmental Change*, 14, 53–67

p. 197 **James Lovelock:** personal communication, 31 March 2006

p. 197 **doubling:** Flannigan, M., et al., 2005: 'Future area burned in Canada', *Climatic Change*, 72, 1–16

p. 198   **early Eocene:** Dawson, M., et al., 1976: 'Paleogene terrestrial vertebrates: northernmost occurrence, Ellesmere Island, Canada', *Science*, 192, 781–2

p. 198   **differences between seasons:** McKenna, M., 1976: 'Eocene paleolatitude, climate and mammals of Ellesmere Island', *Palaeogeography, Palaeoclimatology, Palaeoecology*, 30, 349–62

p. 199   **new mammal species:** Rose, K., 1980: 'Clarkforkian Land-Mammal Age: revised definition, zonation, and tentative intercontinental correlation', *Science*, 208, 744–6

p. 200   **extinction event:** Kennett, J., and Stott, L., 1991: 'Abrupt deep-sea warming, palaeoceanographic changes and benthic extinctions at the end of the Palaeocene', *Nature*, 353, 225–9

p. 201   **Dickens thought:** Dickens, G., et al., 1995: 'Dissociation of oceanic methane hydrate as a cause of the carbon isotope excursion at the end of the Paleocene', *Paleoceanography*, 10, 965–71

p. 201   **Katz found evidence:** Katz, M., et al., 1999: 'The source and fate of massive carbon input during the latest Paleocene Thermal Maximum', *Science*, 286, 1531–3

p. 202   **Geologists have dated:** Storey, M., et al., 2007: 'Paleocene-Eocene Thermal Maximum and the opening of the Northeast Atlantic', *Science*, 316, 587–9

p. 202   **Spain:** Schmitz, B., and Pujalte, V., 2003: 'Sea-level, humidity, and land-erosion records across the initial Eocene thermal maximum from a continental-marine transect in northern Spain', *Geology*, 31, 8, 689–92

p. 202   **'megafan':** Schmitz, B., and Pujalte, V., 2007: 'Abrupt increase in seasonal extreme precipitation at the Paleocene-Eocene boundary', *Geology*, 35, 3, 215–18

p. 202 **the seas acidic:** Zachos, J., et al., 2005: 'Rapid acidification of the ocean during the Paleocene-Eocene Thermal Maximum', *Science*, 308, 1611–15

p. 202 **North America:** Retallack, G., 2005: 'Pedogenic carbonate proxies for amount and seasonality of precipitation in paleosols', *Geology*, 33, 4, 333–6

p. 203 **Rockies:** Sewall, J., and Sloan, L., 2006: 'Come a little bit closer: A high-resolution climate study of the early Paleogene Laramide foreland', *Geology*, 34, 2, 81–4

p. 203 **England and Belgium:** Retallack, G., 2005: 'Pedogenic carbonate proxies for amount and seasonality of precipitation in paleosols', *Geology*, 33, 4, 333–6

p. 203 **own greenhouse effect:** Jahren, A., and Sternberg, L., 2003: 'Humidity estimate for the middle Eocene Arctic rain forest', *Geology*, 31, 5, 463–6

p. 203 **sea temperatures:** Sluijs, A., et al., 2006: 'Subtropical Arctic Ocean temperatures during the Palaeocene/Eocene thermal maximum', *Nature*, 441, 610–13

p. 203 **Air temperatures:** Weijers, J., et al., 2007: 'Warm Arctic continents during the Palaeocene-Eocene Thermal Maximum', *Earth and Planetary Science Letters*, in press

p. 203 **higher rainfall:** Pagani, M., et al., 2006: 'Arctic hydrology during global warming at the Palaeocene/Eocene thermal maximum', *Nature*, 442, 671–5

p. 203 **no ice:** Kerr, R., 2004: 'Signs of a warm, ice-free Arctic', *Science*, 305, 1693

p. 203 **temperatures soared:** Pagani, M., et al., 2006: 'An ancient carbon mystery', *Science*, 314, 1556–7

p. 204 **Gerald Dickens:** Dickens, G., 1999: 'The blast in the past', *Nature*, 401, 752–5

p. 204 **John Higgins and Daniel Schrag:** Higgins, A., and Schrag, D., 2006: 'Beyond methane: Towards a theory for the

Paleocene-Eocene Thermal Maximum', *Earth and Planetary Science Letters*, 245, 523–37

p. 204 **early Eocene:** Lowenstein, T., and Demicco, R., 2006: 'Elevated Eocene atmospheric CO_2 and its subsequent decline', *Science*, 313, 1928

p. 204 **30 times faster:** 2006: 'Lesson from 55 million years ago says climate change could be faster than expected', *Daily Telegraph*, 17 February 2006

p. 204 **searing global heatwave:** Thomas, D., et al., 2002: 'Warming the fuel for the fire: Evidence for the thermal dissociation of methane hydrate during the Paleocene-Eocene thermal maximum', *Geology*, 30, 12, 1067–70

p. 205 **decrease by 85 per cent:** Buffett, B., and Archer, D., 2004: 'Global inventory of methane hydrate: sensitivity to changes in the deep ocean', *Earth and Planetary Science Letters*, 227, 185–99

p. 205 **Arctic Ocean:** see 'Methane hydrates and global warming', RealClimate blog, 12 December 2005, http://www.realclimate.org/index.php?p=227

p. 205 **map:** Hadley Centre, 2005: *Stabilising Climate to Avoid Dangerous Climate Change – a Summary of Relevant Research at the Hadley Centre*, p. 9: 'Increasing natural methane emissions'

p. 207 **tsunami hit the British Isles:** Sissons, J., and Smith, D., 1965: 'Peat bogs in a post-glacial sea and a buried raised beach in the western part of the Carse of Stirling', *Scottish Journal of Geology*, 1, 247–55

p. 207 **Northumberland:** Smith, D., et al., 2004: 'The Holocene Storegga Slide tsunami in the United Kingdom', *Quaternary Science Reviews*, 23, 2291–321

p. 207 **Shetlands:** Bondevik, S., et al., 2004: 'Record-breaking height for 8000-year-old tsunami in the North Atlantic', *Eos*, 84, 31, 289–300

p. 207 **20 metres in the Shetlands:** Ibid.

p. 207 **release of methane hydrates:** Zillmer, M., et al., 2005: 'Imaging and quantification of gas hydrate and free gas at the Storegga slide offshore Norway', *Geophysical Research Letters*, 32, L04308

p. 207 **pockmarks:** see 'Methane hydrates and global warming', RealClimate blog, 12 December 2005, http://www.real climate. org/index.php?p=227

p. 207 **lost any methane hydrate:** Paull, C., et al., 2007: 'Assessing methane release from the colossal Storegga submarine landslide', *Geophysical Research Letters*, 34, L04601

p. 208 **about 10,000 years:** Pagani, M., et al., 2006: 'An ancient carbon mystery', *Science*, 314, 1556–7

p. 208 **evidence from North America:** Wing, S., et al., 2005: 'Transient floral change and rapid global warming at the Paleocene-Eocene boundary', *Science*, 310, 993–6

p. 209 **PETM began dry:** Ibid.

6 SIX DEGREES

p. 217 **Cretaceous period:** Skelton, P. (ed.), 2003: *The Cretaceous World*, Cambridge University Press

p. 219 **Cretaceous rocks:** Ibid.

p. 219 **champsosaurs:** Tarduno, J., et al., 1998: 'Evidence for extreme climatic warmth from Late Cretaceous Arctic vertebrates', *Science*, 282, 2241–4

p. 219 **breadfruit trees:** Jenkyns, H., et al., 2004: 'High temperatures in the Late Cretaceous Arctic Ocean', *Nature*, 432, 888–92

p. 219 **hummocks:** Ito, M., et al., 2001: 'Temporal variation in the wavelength of hummocky cross-stratification: Implications for storm intensity through Mesozoic and Cenozoic', *Geology*, 29, 1, 87–9

p. 219 **North America:** White, T., et al., 2001: 'Middle Cretaceous greenhouse hydrologic cycle of North America', *Geology*, 29, 4, 363–6

p. 219 **tropical Atlantic:** Bice, K., et al., 2006: 'A multiple proxy and model study of Cretaceous upper ocean temperatures and atmospheric CO_2 concentrations', *Palaeoceanography*, 21, PA2002

p. 219 **South Atlantic:** Bice, K., et al., 2003: 'Extreme polar warmth during the Cretaceous greenhouse? Paradox of the late Turonian O18 record at Deep Sea Drilling Project Site 511', *Palaeoceanography*, 18, 2, 1031

p. 220 **adapted to drought:** Skelton, P. (ed.), 2003: *The Cretaceous World*, Cambridge University Press

p. 220 **Antarctic Peninsula:** Poole, I., et al., 2005: 'A multi-proxy approach to determine Antarctic terrestrial palaeoclimate during the Late Cretaceous and Early Tertiary', *Palaeogeography, Palaeoclimatology, Palaeoecology*, 222, 95–121

p. 220 **South Pole:** Brentnall, S., et al., 2005: 'Climatic and ecological determinants of leaf lifespan in polar forests of the high CO_2 Cretaceous "greenhouse" world', *Global Change Biology*, 11, 2177–95

p. 220 **North Pole:** Jenkyns, H., et al., 2004: 'High temperatures in the Late Cretaceous Arctic Ocean', *Nature*, 432, 888–92

p. 220 **Pangaean supercontinent:** Poulsen, C., 2003: 'Did the rifting of the Atlantic Ocean cause the Cretaceous thermal maximum?', *Geology*, 31, 2, 115–18

p. 223 **correlation with warming peaks:** Jenkyns, H., 2003: 'Evidence for rapid climate change in the

Mesozoic-Palaeogene greenhouse world', *Philosophical Transactions of the Royal Society of London A*, 361, 1885–1916

p. 223 **methane hydrate releases:** Kerr, R., 2000: 'Quakes large and small, burps big and old', *Science*, 287, 576–7

p. 224 **algal bloom:** Beckmann, B., et al., 2005: 'Orbital forcing of Cretaceous river discharge in tropical Africa and ocean response', *Nature*, 437, 241–4

p. 225 **vertical rock pipes:** Svensen, H., et al., 2007: 'Hydrothermal venting of greenhouse gases triggering Early Jurassic global warming', *Earth and Planetary Science Letters*, 256, 554–66

p. 225 **pulse:** Ibid.

p. 227 **Most life … wiped out:** Benton, M., 2003: *When Life Nearly Died: The Greatest Mass Extinction of All Time*, Thames & Hudson, p. 168

p. 227 **12 mm of strata:** Kaiho, K., et al., 2006: 'Close-up of the end-Permian mass extinction horizon recorded in the Meishan section, South China: Sedimentary, elemental, and biotic characterization and a negative shift of sulfate sulfur isotope ratio', *Palaeogeography, Palaeoclimatology, Palaeoecology*, 239, 3–4, 396–405

p. 227 **rapid greenhouse warming:** Benton, M., 2003: *When Life Nearly Died: The Greatest Mass Extinction of All Time*, Thames & Hudson, p. 267

p. 227 **soil erosion:** Sephton, M., 2005: 'Catastrophic soil erosion during end-Permian biotic crisis', *Geology*, 33, 12, 941–4

p. 228 **'fungal spike':** Visscher, H., et al., 1996: 'The terminal Paleozoic fungal event: Evidence of terrestrial ecosystem destabilization and collapse', *PNAS*, 93, 5, 2155–8

p. 228 **Karoo Basin:** Smith, R., and Ward, P., 2001: 'Pattern of vertebrate extinctions across an event bed at the

Permian-Triassic boundary in the Karoo Basin of South Africa', *Geology*, 29, 12, 1147–50

p. 228   **fascinating model:** Kidder, D., and Worsley, T., 2004: 'Causes and consequences of extreme Permo-Triassic warming to globally equable climate and relation to the Permo-Triassic extinction and recovery', *Palaeogeography, Palaeoclimatology, Palaeoecology*, 203, 3–4, 207–37

p. 229   **hurricanes:** Ibid.

p. 230   **Siberia:** Kamo, S., et al., 2003: 'Rapid eruption of Siberian flood-volcanic rocks and evidence for coincidence with the Permian-Triassic boundary and mass extinction at 251 Ma', *Earth and Planetary Science Letters*, 214, 75–91

p. 231   **stripped the land:** Benton, M., 2003: *When Life Nearly Died: The Greatest Mass Extinction of All Time*, Thames & Hudson

p. 231   **Oxygen levels:** Weidlich, O., et al., 2003: 'Permian-Triassic boundary interval as a model for forcing marine ecosystem collapse by long-term atmospheric oxygen drop', *Geology*, 31, 11, 961–4

p. 231   **how events unfold:** Ryskin, G., 2003: 'Methane-driven oceanic eruptions and mass extinctions', *Geology*, 31, 9, 741–4

p. 233   **oceanic methane eruption:** Ryskin, G., 2003: Ibid.

p. 233   **sulphurous ocean:** Nielsen, J., and Shen, Y., 2004: 'Evidence for sulfidic deep water during the Late Permian in the East Greenland basin', *Geology*, 32, 12, 1037–40

p. 233   **release of hydrogen sulphide:** Kump, L., et al., 2005: 'Massive release of hydrogen sulfide to the surface ocean and atmosphere during intervals of ocean anoxia', *Geology*, 33, 5, 397–400

p. 233   **DNA mutations:** Visscher, H., et al., 2004: 'Environmental mutagenesis during the end-Permian ecological crisis', *PNAS*, 101, 35, 12952–6

p. 234 **cause of the extinction:** Lamarque, J.-F., et al., 2006: 'Modeling the response to changes in tropospheric methane concentration: Application to the Permian-Triassic boundary', *Palaeoceanography*, 21, PA3006

p. 234 **two agents combined:** Lamarque, J.-F., et al., 2007: 'Role of hydrogen sulfide in a Permian-Triassic boundary ozone collapse', *Geophysical Research Letters*, 34, L02801

p. 238 **wider-scale release:** Bakun, A., and Weeks, S., 2004: 'Greenhouse gas buildup, sardines, submarine eruptions and the possibility of abrupt degradation of intense marine upwelling ecosystems', *Ecology Letters*, 7, 1015–23

p. 238 **olfactory nerve becomes paralysed:** Kump, L., et al., 2005: 'Massive release of hydrogen sulfide to the surface ocean and atmosphere during intervals of ocean anoxia', *Geology*, 33, 5, 397–400

p. 240 **billion years remain:** Franck, S., et al., 2002: 'Long-term evolution of the global carbon cycle: historic minimum of global surface temperature at present', *Tellus B*, 54, 4, 325–43

7 CHOOSING OUR FUTURE

p. 246 **rising four times faster:** Jones, N., 2006: 'Carbon tally shows growing global problem', Nature.com news, http://www. nature.com/news/2006/061106/full/061106-18.html

p. 246 **0.5 [degrees Celsius]:** Teng, H., et al., 2006: 'Twenty-first century climate change commitment from a multi-model ensemble', *Geophysical Research Letters*, 33, L07706

p. 246 **1 degree Celsius:** Wigley, T., 2005: 'The climate change commitment', *Science*, 307, 1766–9

p. 249   **11°C:** Stainforth, D., et al., 2005: 'Uncertainty in predictions of the climate response to rising levels of greenhouse gases', *Nature*, 433, 403–6

p. 249   **'dangerously high':** See http://climateprediction.net/science/pubs/climateprediction_press_release.pdf

p. 249   **already under way:** Wild, M., et al., 2005: 'From dimming to brightening: decadal changes in solar radiation at Earth's surface', *Science*, 308, 847–50

p. 249   **upper extremes:** Andraea, M., et al., 2005: 'Strong present-day aerosol cooling implies a hot future', *Nature*, 435, 1187–90; Bellouin, N., et al., 2005: 'Global estimate of aerosol direct radiative forcing from satellite measurements', *Nature*, 438, 1138–41

p. 249   **models cannot reproduce:** Higgins, A., and Schrag, D., 2006: 'Beyond methane: Towards a theory for the Paleocene-Eocene Thermal Maximum', *Earth and Planetary Science Letters*, 245, 523–37

p. 250   **Cretaceous greenhouse:** Bice, K., et al., 2006: 'A multiple proxy and model study of Cretaceous upper ocean temperatures and atmospheric CO_2 concentrations', *Palaeoceanography*, 21, PA2002

p. 251   **different feedback effects:** Torn, M., and Harte, J., 2006: 'Missing feedbacks, asymmetric uncertainties, and the underestimation of future warming', *Geophysical Research Letters*, 33, L10703

p. 251   **increase climate sensitivity:** Hansen, J., et al., 2007: 'Climate change and trace gases', *Philosophical Transactions of the Royal Society A*, 365, 1925–54

p. 251   **dramatically increase:** Damon Matthews, H., and Keith, D., 2007: 'Carbon-cycle feedbacks increase the likelihood of a warmer future', *Geophysical Research Letters*, 34, L09702

p. 251 **second international study:** Scheffer, M., et al., 2006: 'Positive feedback between global warming and atmospheric CO_2 concentration inferred from past climate change', *Geophysical Research Letters*, 33, L10702

p. 252 **7 per cent chance:** Meinshausen, M., 2006: 'What does a 2°C target mean for greenhouse gas concentrations? A brief analysis based on multi-gas emission pathways and several climate sensitivity uncertainty estimates', in Schellnhuber, H. J. (ed.), *Avoiding Dangerous Climate Change*, Cambridge University Press

p. 253 **politically realistic option:** Stern, N., 2006: *Stern Review: The economics of climate change*, published by HM Govt, available at http://www.hm-treasury.gov.uk/independent_reviews/stern_review_economics_climate_change/sternreview_index.cfm

p. 254 **peak within … *seven years*:** den Elzen, M., and Meinshausen, M., 2006: 'Multi-gas emission pathways for meeting the EU 2°C climate target', in Schellnhuber, H. J. (ed.), *Avoiding Dangerous Climate Change*, Cambridge University Press

p. 258 **IEA projects:** World Energy Outlook 2005 Executive Summary, IEA, 2005, http://www.iea.org/Textbase/npsum/WEO2005SUM.pdf

p. 258 **World Alternative Policy Scenario:** World Energy Outlook 2005 press release, IEA, 2005, http://www.worldenergyoutlook.org/press_rel.asp

p. 260 **figures are startling:** Dukes, J., 2003: 'Burning buried sunshine: Human consumption of ancient solar energy', *Climatic Change*, 61, 31–44

p. 261 **iceberg lettuce:** 2001: *Eating Oil: Food Supply in a Changing Climate*, Sustain/Elm Farm Research Centre,

331

December 2001; summary: http://www.sustainweb.org

p. 261 **consumes ten times more:** Giampietro, M., and Pimentel, D., 1993: 'The tightening conflict: Population, energy use, and the ecology of agriculture', *NPG Forum Series*, October 1993, 1–8

p. 261 **'remarkable level of co-option':** Imhoff, M., et al., 2004: 'Global patterns in human consumption of net primary production', *Nature*, 429, 870–3

p. 264 **'tragedy of the commons':** Hardin, G., 1968: 'The tragedy of the commons', *Science*, 162, 1243–8

p. 264 **focus groups in Switzerland:** Stoll-Kleemann, S., et al., 2001: 'The psychology of denial concerning climate mitigation measures: evidence from Swiss focus groups', *Global Environmental Change*, 11, 107–17

p. 267 **consumed more oil:** Leggett, J., 2005: *Half Gone: Oil, Gas, Hot Air and the Global Energy Crisis*, Portobello Books, p. 59

p. 268 **'these things came together':** 2005: 'Saudi Arabia's Ghawar Field: the elephant of all elephants', *AAPG Explorer*, http://www.aapg.org/explorer/2005/01jan/ghawar.cfm

p. 268 **global economic crash:** Leggett, J., 2005: *Half Gone: Oil, Gas, Hot Air and the Global Energy Crisis*, Portobello Books, pp. 95–6

p. 269 **'powerdown':** Heinberg, R., 2004: *Powerdown: Options and Actions for a Post-Carbon World*, Clairview Books

p. 271 **wedges:** Pacala, S., and Socolow, R., 2004: 'Stabilization wedges: Solving the climate problem for the next 50 years with current technologies', *Science*, 305, 968–72

p. 272 **birds perish:** Marris, E., and Fairless, D., 2007: 'Wind farms' deadly reputation hard to shift', *Nature*, 447, 126

p. 273 **wiped out permanently:** Kempton, W., et al., 2007:

'Large CO_2 reductions via offshore wind power matched to inherent storage in energy end-uses', *Geophysical Research Letters*, 34, L02817

p. 273 **Carbon capture and storage:** IPCC, 2005: *Carbon Dioxide Capture and Storage: Summary for Policymakers and Technical Summary*, WMO/UNEP

p. 274 **save the same:** Gullison, R., et al., 2007: 'Tropical forests and climate policy', *Science*, 316, 985–6

p. 275 **a few hundredths:** Monbiot, G., 2004: 'Fuel for nought', *The Guardian*, 23 November 2004; http://www.guardian.co.uk/ climatechange/story/0,,1357462,00.html

p. 276 **south-east Asian forests:** Ibid.

INDEX

acid rain 171, 228, 230, 249
adaptation 94, 103, 196, 208–9, 220, 235
Afifi, Abdulkader 268
Africa 120; agriculture 89–90, 157–8, 173, 195; ancient 108, 220; disease 151–3; drought 22, 101–5, 173, 194; famine 112, 113; glacial retreat 13–18; monsoon 20–1, 52; rainfall 21–2; refugees 159; refuges 210
Agassiz, Lake 10
agriculture xv, 176, 261; abandonment 8, 174, 211; Africa xv, 89–90, 157–8, 173; Arctic 131; Australia 112, 124, 173; Central and South America 82, 85, 89, 134, 173; China 172–3; decline in 90–1, 157, 172–5, 196; drought-resistant crops 174; Europe 59–60, 62, 89; 'firestick farming' 122; growing season, extended 131, 158, 196; harvest failure 13, 113; India 78–9, 137, 173, 174; intensive 195; irrigation 8, 58, 82, 86, 140, 144, 159, 197; new areas 157, 186–7, 196, 197; North America 5–9, 88–9, 90, 143–4, 158–9, 173; Pakistan 139–41; refuges 210; slash-and-burn 120; UK 89, 210–11; worldwide agricultural drought 173–4
air-conditioning 59, 62, 178
Alaska 25, 221; meltdown 25–6, 75, 131, 187; North Slope fossils 219, 221; rainfall 129, 193

Alexandria, Egypt 163–5, 167
algae 34–5, 37, 58, 224
Algeria 19–20
Alley, Richard 68
Alps xv, 29–31, 58, 150, 177, 180–1, 246
Amazon River 119
Amazonia 32–3, 153, 175; death of 115–21, 190, 209, 252; deforestation 119–20; desertification 194, 209; drought 112, 115, 116, 173
American Association for the Advancement of Science 204
American Geophysical Union 69
Andes 80–5, 108, 115, 119, 238
Angola 104, 105
animals *see* wildlife
Antarctic Ocean 199–200
Antarctica 67, 176; ancient 108–9, 220, 222, 228; coal 222; ice-free 211, 220; ice sheets 64, 71, 146, 167–70
Anthropocene 208, 235
aquifers 8, 53, 158, 166, 170, 194–5
Archer, David 205, 206
Arctic 10, 66, 128–31; agriculture 196; amplifier 24, 76, 189; ancient 109–10, 198–9, 203; ice-free 25–7, 109–10, 130, 170, 186, 203, 220; meltdown 23–8, 66–73, 186–90, 246; peoples 76–7; vegetation 76; wildlife 33, 72–7, 187
Arctic Climate Impact Assessment 75
Arctic Ocean ancient 199, 219, 249–50; freshwater run-off 188; methane

hydrate melt 205–6; sea-ice xx, 72–7, 129, 186–7

Argentina 194

Asian Brown Cloud 136

Atacama Desert, Peru 113

Atlantic Ocean: ancient 201, 218, 219; circulation 9–13, 110, 176; freshwater run-off 10–11; hurricane formation 42–6; North 69; tropical 22

Atlas Mountains, Morocco 180, 181

atmosphere 128–9, 234–5, 236–7; ancient 202–5, 222, 225, 230–1, 249–50; atolls 16, 46–7

Australia 31–8, 95, 211; agriculture 112, 124, 173; bushfires 122–5; coal 221; desertification 194; drought 112, 113, 173, 194

Australian Conservation Foundation 123

avalanches 180; submarine 201, 206–8

Axel Heiberg Island, Canada 203

Baker, Andrew 37

Bakun, Andrew 238

Banda Aceh 207

Bangkok, Thailand 72

Bangladesh 79, 132, 135, 137, 165

Barbados 166

Barrow, Alaska 25

Beever, Dr Erik 40

Bennike, Ole 109

Benton, Michael 230–1

Bhopal, India 238

Bighorn Basin, Wyoming 199

biodiversity 33–4, 91–7, 154–6, 198, 213, 234, 240, 252; Amazon 119; China 171; marine 154; plants 39; under threat 18, 41

biofuels 274–6

Biosphere 2 97

birds 75, 77, 92, 94, 95, 120, 155

Black Sea 237

bogs, thawing 188–9

Bolivia 84, 121

Bombay see Mumbai

Boston, USA 165

Botswana 90, 101–5, 151

Bowman, Malcolm 147

Brahmaputra River 138, 193

Brazil 42, 43, 44, 95, 120, 121, 194; desertification 194; hurricanes 42–4; rainforest 120

Brigham, Lawson 189

British Antarctic Survey 110, 167

British Council 14

British Virgin Islands 38

Broe, Pat 72

Bryden, Professor Harry 11, 12

Buffett, Bruce 205, 206

Bunyard, Peter 119

Burke, Eleanor 22

Burkina Faso 90

Bush, George W. 264

bushfires, Australia 122–5

butterflies 92, 93, 95

Cairo, Egypt 195

calcium carbonate 34, 53–6, 221

Caldeira, Professor Ken 54

California 3–5, 8–9, 85–8, 115, 142, 144–5, 195

California Coastal Range 87

Cameroon 232

Camill, Phil 188, 190

Canada xvii, 131, 188, 196; agriculture 90, 144, 158, 174, 196, 197; ancient 7; arctic 76, 131, 187; forest fires 197; fossil fuels 73, 221, 269; habitable areas 196, 197; rainfall 129; river flows 193, 196

Canberra, Australia 124–5

Cape Floristic Region, South Africa 39

carbon 96, 117, 205, 221–2, 267, capture and storage 271, 273, 276; carbon cycle 56, 116, 117–19, 190, 220–2, 225, 227, 245, 250, 254–5, 256; carbon trading 246, 257; dissolved 188–9; release from seabed 202; release from soil 117–18, 188, 250–1; sequestering 175, 221–2, 223

carbon dioxide xix–xx, 89, 96–7, 252–3; ancient atmosphere 110–12, 203–4, 220, 225, 229, 230, 233, 249–50; and climate sensitivity 248–9; emissions 78, 131, 204, 234–6, 246–7; fertilising effect of 174; from fires 197, 203; ocean acidification 53–6; plant emissions 60; volcanic outgassing 232, 233, 235
Caribbean 38, 113
Carboniferous period 234
Carnegie Institution 54
cars 172, 271, 272, 276
Carson, Rachel 95–6
Cascades, USA 87
Caspian Sea 178
cattle ranching 8, 19, 102, 103, 119, 184
Cayman Islands 63
Central America 82, 83–4, 85, 89, 133–5, 173
Chaco Canyon, New Mexico 5–6
Chad, Lake 20
Chernobyl 176
Chile 89, 194, 211
Chilingarov, Artur 73–4
Chimu civilisation 82, 83
China xxii, 140, 197, 218; coal 221; conflict 197; desertification 197; droughts 51–3; early civilisation 172–3; emissions 257, 258, 264; floods 112; forests 155; hypercapitalism 170–5; monsoon 52–3, 193, 209; water supply 140, 194
civilisations, collapse of 82–3, 131–5, 174–5
climate change xix–xx, 6, 10–11; ancient 15–16, 23, 197–206; conferences on 14; denial 14, 16, 262–5; modelling xv, 105–7, 194–6, 217, 248–51; speed of xxi, 235–6; transient xxi–xxii
climate sensitivity 248–50, 252–3
climate zones, shifting 27–8, 123, 129
climatic envelope 94
clouds 107, 114, 125, 175
coal 221, 222, 226, 234, 262, 269, 271

coasts 72, 76, 145–8, 183, 184–5, 193
coccolithophores 54–5, 56
Collapse: How Societies Choose to Fail or Survive (Diamond) 134
Colorado River, USA 6, 86, 142–5, 167
Columbia 121
Columbia River, USA 87
Comiso, Josefino 75
computer modelling 12, 22; carbon-cycle feedback 116–18; climate xv, 105–7, 194–6, 217, 248–51; El Niño 113–14; global and regional models 106; Hadley Model 39, 59, 105–6, 134; hindcasting 104, 106; hurricane 128; hydrological 139; sea temperature 110
conflict 6, 212–14; over climate refugees 141, 159, 179; over habitable land xxii, 197; over oil 268–9; over water xxii, 85, 86, 141
conservation, site-based 94
continental climates 198
continental drift 218
continental shelf 237; collapse of 201, 206–7
contraction and convergence 257
Cooke, Jennifer 184, 185
cooling: aerosols 135; north-west Europe 9, 12, 211; nuclear winter xviii, 125, 233; Younger Dryas 10
coral: bleaching 34–9, 154–5; reefs 34–9, 42, 63, 91, 154–5, 209, 220, 246
Coral Coast, Fiji 38–9
Cordillera Blanca, Peru 81–2, 84
Cordillera Central, Peru 83–4
Cornwall 28
Costa Rican golden toad 40–1
Cox, Peter 117, 273–4
Cretaceous 56, 200, 217–23, 236, 250
Crump, Marty 40–1
CSIRO 33–4, 122–3
cyclones see hurricanes
Cyprus 62

dams 86, 142–3, 167, 196

Dante xviii, 217, 245–6
Danube River 181
Dasuopu, Tibet 14
Dawson, Mary 197, 198
Democratic Republic of Congo 16
Day After Tomorrow, The 9
deforestation 14, 175–6, 247; Amazonia 119–20; Central America 134
deforestation diesel 276
deglaciation *see* glaciers
denial 14, 16, 262–5
Denmark 73, 149
Department for Environment, Food and Rural Affairs (UK) 78
Department for International Development (UK) 139
desalination plants 178
desert: Amazonia 115, 116; Europe 58, 60, 62, 150, 194; Kalahari Desert 102, 103–4, 105, 194; Marine 224; Mediterranean rim 150–1; North America 3–9; Polar 67; Sahara 21–2, 23, 61, 120, 151, 180, 186, 194, 195, 224; sandstorms 224; spreading 150–1, 186, 194–6, 209, 246
Diamond, Jared 133–4
Dickens, Gerald 200, 204
disease 123, 151–3, 158
drought 60, 193; Africa 101–5; and agriculture 157, 173–4; Amazonia 112, 115, 116, 173; ancient 16, 133–4; Australia 112, 113, 173, 194; Central America 83–4, 133–5; China 51–3; ENSO-related 113; Europe 58, 60, 62, 150, 194; extreme 4, 23; hotspots 173–4; Mediterranean 62–3; Palmer Drought Severity Index 22–3; perennial 194, 209; Sahel 18–19, 22–3; spread of 22–3; threat to woodland 94; UK 177–8, 182; US 3–9, 60, 86–8, 129, 143, 142–5
Doyle, Arthur Conan 153, 154
Dudh Koshi River 80
Dukes, Jeffrey 260, 265–6

dust storms 9, 51–2
Dust Bowl 7–8, 9, 88, 144, 194

Earth: thermal time lag 246, 251; thermoregulation systems 176
earthquakes 207
ecological overshoot 134
economics 170–2
ecosystems 91–7, 175–6, 222–3, 240, 261; Arctic 187; Wet Forest 31–4
Ecuador 84, 89
Eemian interglacial 52, 63, 64
Egypt 19, 195
El Niño 83, 112–15, 224
Ellesmere Island, Nunavut 25, 109, 197–8
Emanuel, Kerry 45, 46
emissions 113, 258, 259; contraction and convergence 257; cuts in 124, 246, 253–9, 276–7; future scenarios 247–8; India 77–8; permits 257; rate of 56, 246, 259; stabilising 276–7; targets 251–9; *see also* greenhouse gases
energy: efficiency 271, 272–3, 275–6; renewable 258, 267–70, 271–7
ENSO 113, 114, 115
Environmental Research Letters 71
Eocene 198, 199, 203–4, 208, 209 *see also* Palaeocene–Eocene Thermal Maximum
erosion: coastal 76, 145–8, 183, 184–5, 193; hillside 134; soil 170, 184–5, 202, 203, 214, 227–8
Estonia 59
ethanol 275
Ethiopia 18, 90, 108, 194, 210
Europe: agriculture 59–60, 62, 89; ancient 218, 220; cooling 9, 10, 11–12; desertification 150, 177–8, 186; drought 58, 60, 62, 150, 194; El Niño effect 113; extinctions 95, 156; floods 148–51, 182–4; heatwaves 57–61, 62, 177–9, 180–2; hurricanes 44–5; rainfall 62, 177; refuges 159; storms 148–51;

temperature rise 175–9, 186; wildfires 62

European Union 256, 276

Everest, Mount 138, 139

extinctions 33, 39–41, 76, 77, 86, 91–7, 208; Anthropocene Mass 235; end-Permian 226–34; Paleocene–Eocene 208–10; human 240; living dead species 156; marine 208–10, 224, 225–6, 229, 233, 234; mass 56, 92, 157, 201, 224, 226–34, 235; plant 39, 76, 77, 91, 93–4, 155, 228

Fahnestock, Mark 69

famine 88, 89, 90–1, 103, 158–9, 213; Africa 112, 113; India 78–9; mass 174; Sahel 18

Faroe Islands 207

feedbacks 190, 252, 255; carbon-cycle 60–1, 116–19, 190, 245, 250, 255; desertification 194; ice-age 255; ice-albedo 28, 70–1; methane 188–90, 202, 204–6

Fiji 38–9, 166

Finland 177

flood barriers 147, 148, 165

flooding: Africa 151; atolls 46–7; coastal cities 145–8, 164–7, 193, 211–12; continental interiors 193, 218; Europe 148–51, 182–4; flash floods 5, 23, 146, 180, 203, 230; monsoon see monsoon; post-ice age 66; storm surge see storm surge; UK xiii–xiv, 148–51, 182–4, 193–4; USA xiv, 115, 145–7, 158, 165–6

food 88–91, 140–1, 166, 172, 213; aid 210; prices 8, 91, 158, 275; production 210, 261–3; shortages 134, 140–1, 158, 174, 186, 275; web, Arctic 75; see also agriculture

Ford, Derek 63

forest 262; ancient 229; boreal 197; carbon emissions 60; deforestation 119–20, 175, 176, 247; die-back 60, 78; montane 17–18; polar 187, 208, 220; reforestation 272; US West 144

forest fires 17, 61–2, 122–5, 197, 203; Amazonia 120–1; Asian 276; Australia 122–5; Europe 58, 59, 61–2; Indonesia 118–19, 121, 136–7; North America 4, 87–8, 144–5

fossil fuels 73, 78, 171–2, 221, 222, 223–6, 261, 267–70

France 58, 62, 149, 177

Francis, Jane 108, 110

freshwater surges 10–11

Frey, Karen 188–9

Friends of the Earth 256

frogs 31–4, 40–1

frost 89, 219

fungi 228

Gabon 90

Gaia Theory 176, 221, 240

Ganges, River 138, 140, 141, 193, 211

gas, natural 7, 73, 222, 260, 261, 268, 269, 271, 272, 276

Gazprom 73

Geology 185

Geology Today 111

Geophysical Research Letters 43, 185

Germany 58, 59, 94, 149, 150, 181, 221

Ghana 18

Giant Sequoia National Park 4

glaciers 10, 80, 110; Alpine 30, 58, 177, 180–1, 187, 246; Andean 80–5; Antarctic 108–9, 167, 168, 169; Arctic 25–6; Greenland 66–9, 138; Himalayas 80; Iceland 130–1; Karakoram 137–42; Kilimanjaro 13–18; Scandinavian 131

Gladwell, Malcolm 23

Glen Canyon Dam 142–3

Global Carbon Project 246

Global Commons Institute 257

global dimming 107, 249

global warming xix; accelerating xiv, 206; ancient xvii; Arctic amplifier 24, 76; climate zone shifts 28; peaks 223–6; positive feedback 189, 252; runaway

204–5, 231, 246, 253, 256, 258; speed of xxi, 235–6; thermal inertia xxi–xxii, 111

Gobi Desert 51, 194, 195

GRACE (Gravity Recovery and Climate Experiment) 70

Great Barrier Reef 34–6

Great Lakes 9

Great Plains 4, 5, 7, 155

Great Depression 210

Greece 177

greenhouse gases xix–xx, 110, 176, 178–9, 188, 236, 247; ocean acidification 53–4; *see also* carbon dioxide; emissions; methane

Greenland 6, 10, 13, 67–70, 76, 129, 187, 252; ancient 24, 75, 109, 219, 233; ice sheet collapse 64–72, 129, 130, 131, 146, 170, 176, 252

Greenpeace 120

Grindelwald, Switzerland 30

Guadalupe River 184

Gulf of Mexico 224

Gulf Stream 9–10, 211, 237

Hadley Centre, UK 22–3, 39, 59, 105–6, 116, 117, 118, 119, 120, 134, 148, 205

Haeberli, Wilfried 30

Hall, T. D. and Billie 184

Hamilton, Dr Gordon 69

Hansen, James 27, 65–6, 67, 70, 71–2, 115, 166

Harappan civilisation 174, 175

Hardin, Garrett 264

Harrison, Gary 76–7

Harvard University 204

Hawaiian 92

Hayward, Dr Alan 110, 111

heat 61, 122–3, 193, 197, 209–10, 231

heatstroke 57–8, 62

heatwaves 6, 186, 197, 202; Alpine 30, 31, 177; Australia 173; Europe 57–63, 150, 178–9, 202; winter 180

Helheim Glacier, Greenland 68, 69

Higgins, Craig 29

Higgins, John 204

High Tide xiii–ix, xv, 46–7, 77, 81

Hilbert, David 33–4

Hill, Robert 108

Himalayas 80, 108, 138, 173

Hoegh-Guldberg, Ove 35–6, 37, 38

Hoerling, Martin 21, 102–3, 104

Holland, Marika 26–7

Holocene 20, 21, 24, 66, 107

Hong Kong 171

housing 183, 272, 276

Houston, Texas 125–8

Howat, Ian 67, 68, 69

Huntingford, Chris 273–4

hurricanes 42–6, 125–8, 129, 146; ancient 219, 229–30; Catarina 42–3; Europe 44–5, 149, 185; Floyd 146; formation 42–6; Galveston 1900 126; hypercanes 230; Katrina xiv, 38, 42, 46, 126, 166; modelling 106; Odessa 126, 127; sea temperature and 126, 229–30; storm surges xiv, 145–9, 165, 182; strong 125–8, 219, 230; Rita 46; Vince 44; Wilma 46

hydroelectricity 17–18, 58, 62, 84–5, 87, 140, 178, 181

hydrogen sulphide 233, 237

hydrological cycle 224

ice ages xvii–xviii, 6, 9, 10, 24, 135; and El Niño 114; modelling 106, 251

ice-albedo feedback 28, 70–1

ice caps 26–7, 64–72, 81, 130–1, 197, 208, 220, 246

ice cores 6, 14, 15–16, 64, 81

ice sheets 64–72, 129, 130, 131, 146, 166, 167–70, 176, 193

ice shelves 168–9

icebergs 68, 113

Iceland 130–1

Inconvenient Truth, An 263

India xxii, 77–80, 135–7, 173; agricultural 78–9, 137, 173; ancient

218; drought 173; Environment
Ministry 78; famine 78–9; monsoon
21, 52, 79, 135–7, 173, 209, 219; water
table 173; warming 102, 104
Indian Ocean 136
Indonesia xvii, 118–19, 121, 136–7, 206,
211, 276
Indus, River 137–42, 174
Industrial Revolution 112
Inter-American Commission on Human
Rights 77
interglacial, Eemian 52, 63–6, 107
Intergovernmental Panel on Climate
Change (IPCC) xiv, xx–xxi, 13, 24, 65,
66, 70, 71, 72, 77, 118, 130, 158, 173,
217, 225, 246, 251, 256
International Energy Agency (IEA) 258
intertropical convergence zone 151, 193
International Rice Research Institute 157
Inuit peoples 76, 77
irrigation 8, 58, 82, 86, 140, 144, 158, 196
Italy 44, 63

Jacobshavn Isbrae glacier, Greenland 68
James Cook University 33
Japan 194
Jequetepeque River 82, 83
jet streams 28
Jones, Brian 63
Jones, Chris 273–4
Joshi, S. K. 78
Journal of Climate 128
Jurassic Period 224–5, 226, 234

Kalahari Desert 102, 103–4, 105, 194
Kangerdlugssuaq Glacier, Greenland 69
Karakoram range 137–8, 139, 173
Karoo plain, South Africa 225, 228
Katz, Miriam 201, 206
Keipper, Dr Vince 13–14
Kennett, James 199–200
Kenya 13, 151
Khama, Sir Seretse 102
Kidder, David 228–9

Kilimanjaro, Mount 13–18, 81
Kiribati 47
Knutson, Tom 128
Korea 186, 194
Krakatoa 201–2
Kruger National Park, South Africa 95
K2 137, 138, 139
Kyoto Protocol 246, 258

lakes: ancient 16, 22; glacial 80, 130;
pollution 171; saline 4–5
landslides 31; submarine 201, 206–8,
211–12
Lawrence Berkeley National Laboratory
145
Lebensraum xxii, 210
Lesotho 90
Libya 19, 44
Lima, Peru 81, 83–5, 114, 195
localism 210, 247, 278
logging, illegal 120
Loire River, France 58
London 9, 12, 72, 148, 165
Lovelock, James 176, 197, 210, 214, 221,
240
Lowe, Jason 148
Luxembourg 61

Madagascar 194, 218
Madeira Islands 44
maize xv, 89
malaria 151–3
Maldives 47
Mali 90
Manabe, Syukuro 194, 196
Marshall Islands 47
Matterhorn 29–30, 58
Mayans 131–5, 174, 175
Mead, Lake 142
Medieval Warm Period 5–7
Mediterranean 44–5, 61–3, 150, 178, 179,
182
Mega-Chad, Lake 20
Meghna delta 211

Meltwater Pulse 1a flood 66
Meridional Overturning Circulation (MOC) 9–10, 11, 13
Meshian quarry, China 226–7
Meteorological Office (UK) xiv, 22–3, 59, 148
methane xix, 188–90, 201, 202, 225, 250, 252; methane-air clouds 232–4, 238
methane hydrate 200–6, 211, 223–4, 225, 231–4, 238, 250, 253, 270
Mexico 95, 131, 159; Gulf of 224, 237
migration: human 85, 124, 134–5, 141, 173, 175, 181–2, 199, 211; plants 93–4, 208–9; species 93–4, 154, 208–9
Mississippi River 224
Moche civilisation 82, 83, 113
modelling *see* computer modelling
monsoon 20–1, 209, 227; acid 230–1; African 21, 52; Asian 21, 79, 135–7, 173, 209, 219; Australian 123; China 52–3, 193, 209; disruption 135–7; intensification 136–7, 173; North America 8, 203
Mont Blanc 30, 180
Monteverde Cloud Forest, Costa Rica 40, 92
Moon Lake 4–5
mountains 30–1, 85, 86, 179–81; deglaciation 137–42, 180; formation 218, 229; permafrost melt 30–1; refuges 211, 213
Mozambique 151
mudflows/mudslides 80, 115, 134, 180
Mumbai, India 72, 135–7, 165
Murray-Darling River basin, Australia 124

Namibia 104, 105, 237
NASA 27, 65, 70, 75, 166, 168
National Biodiversity Institute (South Africa) 39
National Center for Atmospheric Research (US) 26–7, 118
National Oceanic and Atmospheric

Administration (US) 102–3
National Snow and Ice Data Center (US) 27
Native Americans 5–6
Nature 37, 45, 64, 67, 94, 115–16, 150, 156, 204
nature reserves 94, 235
Nebraska 6–7, 88, 246
Negev Desert, Israel 228
Nepal 80, 137
Netherlands 149, 150, 152
Nevado Huascarán 14, 82
New Orleans xiv, 42, 126, 128, 147, 165, 213
New York City xvii, 9, 145–7, 165
New Zealand 211, 222
Nigeria 18
Nile, River 17, 19, 195, 211
Nile Delta 163–5
nomads, climate 159
North Africa 16, 18, 20, 21, 61, 218
North America: ancient 197–8, 218, 219; desertification 194; drought 5–9, 173; extinctions 39–42; flooding 218, 219; Medieval Warm Period 5–7 *see also* Canada; USA
North Sea 149, 207
Norway 10, 73, 131, 207
nuclear power 270, 271, 272
nuclear weapons 141, 212, 233, 272
nuclear winter xviii, 125, 233, 272
Nyos Lake, Cameroon 232

oceans; acidification 53–6, 203; ancient 108, 110, 199–203, 219, 220, 223, 229–30, 233; anoxic 223–4, 227, 229, 236, 237, 267; carbon sequestering 175; circulations 9–13, 108, 109–10, 237; deep-sea extinction 199–200; freshwater surges 10–11; methane 190; oscillations 112; pollution 176, 224; salinity 229, stratification 224, 237; sulphurous 233; temperature and hurricanes 42, 43, 44–6, 229–30;

warming process 222; warming spikes 223; *see also* sea levels
Oechel, Walter 189
Office of Science and Technology (UK) 183
oil 73, 78, 171–2, 222, 223–6, 261, 267–70
Oklahoma 7
Oregon 85, 87
Oxford University xv, 104, 248
oxygen 231, 233; isotopes 15, 227
ozone layer 233, 238

Pacala, Steve 270–1, 271–2
Pachauri, Rajendra 77–8
Pacific Islands 46, 194
Pacific Ocean 10, 46, 54, 114, 115
Pakistan xxii, 80, 137, 139–42, 209
Palaeocene 199, 200, 202, 203–4, 225
Palaeocene–Eocene Thermal Maximum (PETM) 203–4, 205, 208–9, 223, 236, 249
palaeoclimate 52, 107, 166, 249, 251
Papua New Guinea 113
Paris 57, 58
Parizek, Byron 68
Patagonia 194, 211
peat 202, 221
Penn State University 68
permafrost xvii, 24, 25, 58, 76, 187, 188, 252; Alpine 30–1; methane release 187–8, 252, 253
Permian–Triassic boundary 225–34, 235, 237, 238, 239, 252
Peru 14–15, 80–5, 112–13, 114, 224
Pezza, Alexandre 43
pikas 39–40, 42, 91
Pinatubo, Mount 125
plankton 54–6, 75, 175, 222, 224, 267
plants 180–1; ancient 218, 219, 220; Cape Floristic Region 39; carbon emissions 60; extinction 39, 76, 77, 91, 93–4, 155, 228; migration 93–4, 208–9; tepui 153, 154; Wet Tropics 32

Pliocene 105–12, 114, 246, 249
Po River, Italy 58
polar bears 27, 72–4, 75, 156
polar regions 108–10; ancient 220, 229; belt of habitability 209; deserts 67; ice-sheet collapse 64–72, 129, 130, 131, 146
politics 159, 210, 247, 254, 271
pollution 171, 176, 178–9, 224, 249
Pontresina, Switzerland 30
population: growth 5, 92, 152, 171, 214, 247; movements 93–4, 124, 173, 175, 181–2, 211; prehistoric crash xviii; reduction in 5, 214
Portugal 44, 58, 61, 177, 178, 186
positive feedbacks *see* feedbacks
Powell, Lake 143
Proceedings of the National Academy of Sciences 87
protea flowers 39, 42, 91
Pueblo Indians 5

Queensland, University of 35
Queensland Wet Tropics rainforest 31–4, 91, 95
Quelccaya ice cap 81

rainfall 17, 101–3; acid rain 171, 228, 230, 249; Africa 21–2, 101–3; ancient 203, 220; Australia 123–4; Central and South America 119, 133–5, 209; convective rainfall 185; Europe xiii, 62, 149, 150, 185; flash floods 23; hurricane xiv, 126; increased 21, 185–6, 194; North America 129, 185; reduced 8, 62; seasonality 62; soil erosion 184–5, 203; UK xiii–ix, 185, 211; *see also* monsoon
rainforest: Amazon 115–21, 153, 175; ancient 203, 220; fires 193; habitat loss 31–4; Queensland Wet Tropics 31–4
Rapley, Chris 167
refugees xiv, 79, 85, 126, 159, 197, 211; conflicts over 141, 159, 197; from

coastal areas 165–6; from Mediterranean 180
refuges 179, 181–2, 210–11
reservoirs 85–6, 87, 124, 133, 140
Revenge of Gaia, The (Lovelock) 176
Rhine River, Germany 58, 94, 150, 181
Richardson, Katherine 56
Rimac River, Peru 81, 84, 195
Rio Santa, Peru 81–2
rivers: flows 58, 193–4, 195, 196; Indus 137–42; USA 142–5
rockfalls 29–30, 31, 58
Rocky Mountains 86, 87, 144, 145, 155, 181, 203
Royal Society 54
Rumsfeld, Donald 247
Russia 73, 89, 174, 177, 178, 196, 197, 221
Rwenzori Mountains, Uganda 16–17
Ryskin, Gregory 232–3

Safsaf Oasis, Egypt 19
Sahara 61, 120, 180, 195, 224; expansion of 151, 186, 194; greening 21–2, 23; wet and dry cycles 18–21
Sahel 16, 18–23, 194
salinity 5, 11, 124, 229
Salt Lake City, Utah 86
San Joaquin River 86, 142
sand dunes 7–8, 18, 19, 82, 121; remobilisation 104; stabilised dunefields 103–4
Sand Hills, Nebraska 7
sandstorms 8
Sardinia 62
Saskatchewan, Canada 7
Saudi Arabia 195, 268
Saunders, Victor 29
Scambos, Ted 27
Schrag, Daniel 204
Science 12, 69, 144, 201
Scientific American 65
Scotland 28, 105–6, 182, 184, 207
sea ice xx, 10; melt 12, 26, 27, 28, 74–5, 129, 186–7

sea levels: ancient 63–5, 109, 218, 229, 246; rising 46–7, 65–72, 131, 146–7, 163–70, 193, 211, 252
seals 27, 75, 77
SEDAPAL 83–4
Seward Peninsula, Alaska 25
Shanghai, China 112, 128, 165
Shea, Dennis 46
Sheehy, John 157
Shelley, Percy Bysshe 174–5, 177
shellfish 54
Shetlands 207
Shishmaref, Alaska 77
Siachen Glacier 138
Siberia xxii, 11, 13, 25, 155, 188–9, 193, 196, 197, 220, 230, 246; ancient 219, 220; drought 173; newly habitable areas 196; permafrost melt 187, 252; rivers 11, 193; volcanoes 230
Sierra Nevada, California 3, 86, 87, 145
Simmonds, Ian 43
Smith, Larry 188–9
smog and smoke 121, 122, 125
snow 67, 86, 119, 178, 181, 183–4, 187; Alps 180, 181; Arctic 187; disappearance of 194; Europe 178; snowmelt feedback 27; snowpack 85–6, 87, 88, 143, 195; UK 183
society 263, 266; collapse of 141, 159, 174; localism 210, 247, 278
Socolow, Robert 270–1, 271–2
soil 175–6, 197; carbon release 117, 188, 250, 251; desiccation 18, 150, 187, 194, 195; erosion 170, 184–5, 202, 203, 214, 227–8; thaw and methane release 188–9
solar power 270, 273, 276
solar radiation 75, 240
Somalia 18
South Africa 39, 91, 95, 104, 173, 225, 269
South America 10; agriculture 89, 173; ancient 218, 220; desertification 194, 209

Southern Nevada Water Authority 86
Spain 44, 62–3, 177, 178, 186, 202
spring, early 92
Stanley, Mount 17
Stern, Sir Nicholas 253
St Helens, Mount 201–2
Storegga Slide 207
storm surges xiii, 146, 148–9, 165, 182
storms 28, 113, 124, 177: Europe 148–51;
    intensity index 45; tempestites 219; UK
    xiii–xiv, 181–4 see also hurricanes
Stott, Lowell 199–200
subtropics 134, 158, 209, 220
Sudan 19
sulphate aerosols 107, 135–6, 249
summer, lengthening 62
survivalism 212–14
Swaziland 90
Sweden 131, 177
Switzerland 30, 177, 181, 264
Sydney, Australia 122, 185–6
synfuels 269

Tanzania xv, 13, 18, 151
Tasmania 123, 124, 173, 211
temperature: ancient 64, 218–19,
    249–50; rising xx–xxi, 28, 120, 193,
    227; sea 126, 229–30; speed of rise 93
    see also heat; heatwaves
tempestites 219
tepuis 153–4
Texas 88, 125–8, 184–6
Thailand 263
Thames, River xiii, 12, 148
thermal inertia xxi–xxii, 111
Thomas, Chris 94–5, 156
Thomas, David 104, 105
Thomas, Dylan 240–1
Thomas, Jeremy 93
Thompson, Lonnie 14–15, 80–1
Tibet 14, 137, 141
Tierra del Fuego, Chile 211
Tikal, Mexico 131–2, 133
Titanic 113

tornadoes 125
tourism 18, 61, 62, 164, 263
trade winds 114, 151, 193
tragedy of the commons 264
Transantarctic Mountains 108–9, 169,
    228
transport 171–2, 271–2, 275–6, 277–9
trees 93–4; leaf stomata 110–11; stumps
    3–4, 203, 218–19; treeline 109, 144
Trenberth, Kevin 46
tropical storms see hurricanes
tropics 159, 193, 208; agriculture 157–8;
    covergence zone 151, 193
Truffer, Martin 69
Trujillo, Peru 82–3
tsunamis 206–8, 212
Tuleya, Bob 128
tundra xvii, 10, 109, 174; shrub
    encroachment 24, 75, 76, 187
Turkey 62, 177
Tuvalu 46–7

Uganda 16–17
Ukraine 177
UNESCO World Heritage Sites 32, 34, 39
uninhabitable zones 195, 197, 209
United Kingdom 108, 181–4; agriculture
    89, 210–11; cooling 12, 211; drought
    177–8, 182; emissions targets 257;
    flooding xiii–xiv, 148–51, 182–4, 193;
    fuel protests 259–60; heatwaves 57,
    177–8; land loss 170; as refuge 179,
    181–2, 210–11; storms xiii–xiv;
    tsunami 207; weather patterns 27–8,
    93, 210–11
United Nations (UN): Climate Change
    Conferences 14, 76–7; Kyoto Protocol
    246, 258; Millennium Ecosystem
    Assessment 92
United States 77, 195, 220; agriculture
    5–9, 88–9, 89, 90, 143–4, 158–9;
    Colorado River 6, 86, 142–5, 167;
    conflict 197; desertification 194;
    drought 3–9, 60, 87, 129, 143, 144–5;

emissions 257, 258, 260, 264, 274;
extinctions 39–40; flooding xiv, 115,
145–7, 158, 165–6; fossil fuels 73, 221;
hurricanes xiv, 38, 42, 46, 125, 166;
north-eastern 113; north-western 129;
refugees 159; refuges 159; soil erosion
184–5, 186; southern 6, 8; water crisis
85–8, 195; western 8, 60, 143–5;
wildfires 4, 87–8 *see also* North
America
USDA Forest Service 145

vacuum bombs 232
Vatnajökull, Iceland 130–1
Venezuela 121, 153
Venice 165
Vezina, Jennifer 63
Virgil xviii, 217
volcanoes xvii–xviii, 14, 130, 201–2, 220,
227, 232

Wales 28
Walker River 3
walruses 75
war *see* conflict
warming spikes 223
Washington State 85, 87
water xiv, 80–2; conflict over xxii, 85, 86,
140; 'fossil' supplies 140; and glacial
retreat 17; and montane forest 17;
rationing 124; saline penetration 124,
166; shortages 53, 58, 63, 81–8, 123–4,
129, 137–42, 196; water tables 196;
water transfer project 53; waterholes
17
Watt-Cloutier, Sheila 77
Weddell Sea 199
wedge approach 270–9

Weeks, Scarla 238
wells 133, 195
West, Robert 197, 198
wetlands 86, 102
Wet Tropics rainforest 33–4, 91, 95
wheat 78, 89, 124, 157
When Life Nearly Died (Benton) 230–1
wildfires *see* bushfires, forest fires
wildlife 91–7, 213; Amazonia 120;
ancient distribution 197–9; Arctic 25,
33, 72–7, 187; coral reefs 34–5; habitat
loss 32–3; marine 209–10; tepuis
153–4; Wet Tropics 32–3 *see also*
extinction
Williams, Dr Steve 33
Williams, John 155
Wilson, Edward O. 95, 96
wind 104–5, 123, 150, 182; trade winds
114, 151, 193
wind turbines 270, 271–2, 277
winter 6, 12, 181, 196
World Wildlife Fund 39
Worsley, Thomas 228–9

Xiaoping, Deng 171

Yakima Valley, USA 143–4
Yangtze River 53, 112, 138, 167, 193, 211
Yellow River, China 171, 193
Yellowstone National Park 5
Yemen 195
Yosemite National Park 4
Younger Dryas 10

Zachos, Jim 204
Zambia 104, 105
Zimbabwe 90, 104, 105, 151